Science Anxiety

Fear Of Science
And How To Overcome It

Jeffry V. Mallow

Professor of Physics
and Dean of Mathematics and Natural Science
Loyola University of Chicago

Foreword by Sheila Tobias

H&H Publishing Co., Inc.

55638

Q♠

181

M175

1986

H&H Publishing Co., Inc.
2165 Sunnydale Blvd., Suite N
Clearwater, FL 33575

ISBN 0-943202-18-3

Printing is the last number.

10 9 8 7 6 5 4 3 1

To David
with love

Contents

Foreword

Of the many myths about science that are explored in this book, perhaps the most pervasive is the idea that the learning and doing of science (and mathematics) is purely a matter of the mind. There is, to be sure, an intellectual rigor required of those who would master science, but as Jeffry Mallow points out in the pages that follow, emotions cannot be left outside of the classroom or laboratory along with the students' coats and boots.

When positive, emotions can contribute to the pleasure and excitement of doing science. When negative, they can intrude on the ability and willingness to learn. In short, emotions ranging from aesthetic satisfaction and the excitement of discovery to anxiety, fear, and frustration play a very important role in determining who succeeds at science and why.

Science anxiety is no simple phenomenon. It cannot be eliminated merely by encouraging and reassuring students who exhibit it. Anxiety causes science avoidance which, in turn, contributes to the poor mastery of certain basic skills. It may intrude on the development of formal reasoning powers: eroding attention to detail, muddling thinking, and causing lapses in memory. So long as the anxiety is not acknowledged and dealt with, the student has no choice but to conclude that he or she is "dumb in science," and this conclusion inevitably causes failure and unwillingness to continue to learn.

In our technologically complex society science avoidance is disastrous both for those individuals attempting to move ahead vocationally and for those who need to have an understanding of science in order to make policy. A recent study pointed out that while America produces well trained scientists and engineers in sufficient number, "science illiteracy" among the population at large reduces productivity. In Japan more than half the leadership in the private and public sectors has had a science or engineering education. In this country, most of the elected officials (especially lawyers) who support and implement policies are inadequately schooled in science.

There are two ways to attempt to cure this national ailment. One is to revise the curriculum at high school and college levels and to require more science and mathematics of everyone. The other is the approach that Jeffry Mallow recommends: to confront and to deal directly with science avoidance and science anxiety.

Who has responsibility for the cure? Just as the problem is complex, the solutions are multifaceted as well. Mallow covers a wide range of possibilities: teaching sub-basic skills, like how to read a science textbook; slowing up the curriculum for those who need a longer time to grasp the concepts; challenging the science teacher to resist the temptation to teach science as if it were an elite subject for an elite population. Better diagnostics could be used to predict which students are ready for formal reasoning and which still need to approach the subject through concrete materials.

Regardless of their own science training, parents have an obligation to encourage their children, and most especially their daughters, to take as much science as they can.

But in the end Mallow seems to argue that the victim has to come to grips with his or her own assumptions, attitudes, and behavior toward science. In place of remedial efforts, Mallow would have us teach coping skills, either as part of a formal Science Anxiety Clinic, such as the one he has operated at Loyola University for several years, or by individual effort. Much of this book outlines the steps that parents, teachers, victims, and psychologists can use in dealing with science anxiety. This book thereby represents a major contribution to our thinking about the cure.

Sheila Tobias
Author, *Overcoming Math Anxiety*
(W.W. Norton, 1978; Houghton Mifflin, 1980.)

Preface

In the last few years, a spate of reports has emerged assessing and criticizing the preparations of elementary, secondary, and post-secondary students in the sciences and mathematics. The best known of these is probably the report of the National Commission on Excellence in Education entitled "A Nation At Risk." It decries American education's "rising tide of mediocrity," specifically warning that "individuals in our society who do not possess the levels of skill, literacy, and training essential to this new era will be effectively disenfranchised." The report quotes science educator Paul de Hart Hurd: "We are raising a new generation of Americans that is scientifically and technically illiterate." It reiterates former National Science Foundation Director John Slaughter's warning of "a growing chasm between a small scientific and technological elite and a citizenry ill-informed, indeed uninformed on issues with a science component."

The consequences of scientific and technical mediocrity for the nation in its position vis a vis other nations have been widely discussed in many reports and articles. "A Nation At Risk" states the problem succinctly and clearly:

> . . .Knowledge, learning, information, and
> skilled intelligence are the new raw materials of
> international commerce. . . For our country to
> function, citizens must be able to reach some
> common understandings on complex issues,
> often on short notice and on the basis of con-
> flicting or incomplete evidence.

Preface

While the problem of science illiteracy has a number of causes, one of the most obvious is widespread "science anxiety" -- fear of science. This includes not only fear of the consequences of science, pollution, reactors, weapons, but the fear that one cannot learn, cannot grasp, even rudimentary scientific ideas. In fact, the National Research Council in its 1982 report, "Science for Non-Specialists: The College Years," explicitly states that "College science education should enable nonspecialists to overcome fears that might prevent them from launching a lifetime learning experience about science and technology."

This is a book about fear of science and how to overcome it. It is written for students, parents, teachers, and interested readers in general. My goals are to examine the causes of science anxiety and to investigate its consequences, both personal (what jobs require some knowledge of science?) and political (how do we decide if nuclear reactors are safe or not?). Among the topics to be explored are: How people are taught to fear science; how lack of science education acts as a "job filter," discriminating against women and some minorities; how science is artificially separated from the liberal arts; how science anxiety is similar to and different from math anxiety; and how fear of science may be overcome. The method that we have used to reduce science anxiety among students in our Science Anxiety Clinic at Loyola University of Chicago will be discussed in detail, so the science-anxious reader can discover which techniques are best suited to his or her needs. Interested science teachers and counselors will find information about setting up a science anxiety clinic at their own institution. And readers who are neither students nor teachers, but who wish they could understand science, will discover that they can once they overcome their fears.

This does not mean that anyone who reads this book will become a scientist, just as not everyone who reads a book on art goes on to become a painter. It does mean that science, like art, should be part of an educated person's repertoire and that an appreciation of science, like art, enhances and enriches our lives.

Jeffry V. Mallow
Loyola University of Chicago
December 1985

Acknowledgments

I should like to thank the following people for their help in developing the Science Anxiety Clinic at Loyola University of Chicago and for providing me with many of the ideas that appear in this book:

Dr. Marian Grace, staff psychologist at the Loyola Counseling Center and cofounder of the Science Anxiety Clinic, who also provided useful criticism of sections of this book; Dr. Daniel Barnes, director of the Counseling Center; Dr. Richard Bukrey of Loyola's Physics Department; and Drs. Rosemarie Alvaro and Joseph Hermes for their research in science anxiety. Dr. Alvaro developed much of the wording in the Clinic sessions, discussed in Chapter 6; she and Dr. Grace developed the Clinic procedures and the Science Questionnaire (Chapter 3).

I am grateful for a Mellon Foundation Grant, which provided the initial funding for the Clinic. I thank the various administrators at Loyola, who helped me bring the Clinic and this book to fruition.

I thank my mother, Mrs. Edith Lerner, who read this manuscript in its early stages and suggested many improvements, especially in areas where I lapsed into science jargon or stated ideas unclearly.

I am grateful to my sister, Mrs. Brona Neville-Neil, for her hospitality while I was writing this book.

Finally, I thank Professor Arthur J. Freeman of the Physics Department of Northwestern University, who taught me to love science.

1
What Is Science Anxiety?

Turn on your television some Saturday afternoon. Find the channel with the monster movie. Now settle back in your chair and watch the stereotypes parade across the screen. There goes the beautiful girl. Not far behind her, and gaining quickly, is the monster. And who is that lurking in the shadows: the guy in the lab coat, with the rimless spectacles and the crazed look on his face? It's the scientist, of course. Brilliant, cold-blooded, seeking to cure humanity's ills but instead unleashing new horrors on the human race, he inspires respect and terror in us at the same time. What he does *not* inspire is the desire to emulate him. And that is the root of science anxiety. From early childhood, most of us are taught that we cannot grasp science. That's for someone else: the brain, the oddball, the misfit. As soon as we encounter science in school, we learn to tell one another that it is too difficult for us. We absorb the heritage of science anxiety even as we study the concepts of science. As we grow up and grow older, our avoidance of science increases. We suffer through required courses in biology, chemistry, and physics in high school or college. Then we put the whole dismal experience behind us forever. We come to view science as a necessary component of our lives but certainly as something beyond our ability to comprehend.

How does science anxiety manifest itself? Among students there is widespread avoidance of science courses in high school and college. As soon as they are given the choice (usually in their first year of high school), they opt out of as many science courses as possible. Enrollments in these courses across the country are well below enrollments in nonscience courses. Only those courses that are absolutely required by the school are well attended. And even in these classes, science anxiety rears its ugly head. Some students avoid asking questions of the instructor. Others panic on exams. Some stare at the test questions for the entire exam period without writing anything down. Many students attempt to memorize the material of science as if it were a poem, rather than trying to grasp and apply the concepts. Even if they get through with passing grades, they will always regard the study of science as one of the most difficult and unpleasant experiences of their lives.

Science anxiety carries over into society at large. Antiscience attitudes are quite the fashion. Writers, artists, and teachers of humanities boast of their ignorance of all things scientific. Martin Green, an English professor who does not fit this antiscience mold, relates an interesting story[1]:

> Dennis Flanagan, editor of *Scientific American*,
> described a two-cultures clash between himself and
> Pauline Kael, the *New Yorker's* film critic.
>
> Immediately upon their introduction, Kael apparently
> said, with what sounded to Flanagan like either
> complacency or aggression, "I know nothing whatever
> about science."
>
> "Whatever became of the idea that an educated person
> should know a little something about everything?"
> Flanagan replied.
>
> And Kael (her eyes snapping a bit, as Flanagan
> remembers) exclaimed, "Ah, a Renaissance hack."

The "two cultures" to which Green refers are the humanities and the sciences; the term "two cultures" was invented by C.P. Snow in a famous essay in which he analyzed the complete lack of communication between these two groups.[2] The gap has not lessened since Snow wrote his essay more than 25 years ago. Universities build separate buildings for sciences and humanities. The rationalization for this is probably efficiency, but in truth the two groups of scholars are not comfortable in each other's company.

Other examples of science anxiety in our society (and others) are not difficult to find. Cartoons depicting scientists are similar in style to monster movies. Popular accounts of the greatest scientist of our, and perhaps all, time -- Albert Einstein -- invariably emphasize his eccentricity, as if this must go hand-in-hand with his achievements. In point of fact, Einstein was not especially eccentric. His casual dress, although unorthodox a quarter century ago, would go unremarked upon today. He was a good athlete and an accomplished sailor; his interests included music, politics, and religion in addition to physics. He was in fact a fine role model, with many qualities that young people could emulate -- but this side of him has been buried by media sterotypes of scientists.

Reports of scientific events on the evening news evoke a science-anxious response in newscasters. After a report of the latest pictures of Saturn, or the latest breakthrough in biological research, you can usually count on one of the newscasters to say with a giggle: "You understand that? Not me!" We learn to accept these nervous jokes about the incomprehensibility of science. We learn not to learn science. And we pay for this attitude dearly. We pay first by limiting our options for meaningful careers. If we shy away from studying science when we are in school, then we slam shut in our own faces the door to many rewarding jobs. Medical and dental schools typically require entering students to have completed, with high grades, two years of biology, two years of chemistry, one year of physics, and one year of science's sister subject, mathematics. Innumerable jobs in industry, both blue- and white-collar, require a certain amount of technical training. Even if this is not apparent in the first few years of a person's career, it later becomes clear that those workers with a technical background and with the initiative to acquire further technical skills are promoted first and fastest. The rapidly growing health professions, including nursing and dental hygiene, are requiring the study of more and more science. The student who, because of science anxiety, avoids taking science courses in high school or college severely limits his or her options for the future.

This aspect of science as a career filter operates against all students, but proportionately more against women and disadvantaged minorities. The percentage of these groups in the scientific professions, and indeed in any fields that require even a modicum of training in science, is very low. If the first dogma of a science-anxious society is "almost nobody can do science," then the second is "and certainly women and disadvantaged minorities can't!" This is not to say that such messages are always overt (although some black and Hispanic students in our Science Anxiety Clinic have related that they were told directly by teachers, "You people can't do science"); nevertheless, children learn quickly what is an appropriate role for them.

Much has been written about how girls are turned away from math.[3] The same is true for science, and probably to a greater degree. If we chart the number of girls and young women in science courses in high school and college, then we quickly see that the higher the grade level, the lower the proportion of females. If typically half of first-year college students majoring in biology are women, this percentage drops below one-third by the senior year. Women physics majors are a rarity indeed. In my experience at universities, I have personally observed little blatant discrimination against women (although some surely exists). However, by the time they enter college, many women, even if they like science, have internalized earlier sexist messages and are therefore convinced they cannot succeed. And there's no prophecy more powerful than a self-fulfilling prophecy. The scarcity of female scientists as role models is itself a clear message to girls: "Stay away!"

Another consequence of science anxiety is widespread avoidance of anything quantitative. Here, science anxiety and math anxiety are similar.

Science is quantitative. It is not enough to say, "Hydrogen combines with oxygen to make water"; we must know how much hydrogen and how much oxygen and how fast the reaction occurs. It is not enough to say "Uranium 235 is radioactive"; we must know how much radioactivity is emitted and how much energy it has and what that number 235 means.

As science enrollments have dropped, some science teachers have unfortunately tried to reverse this trend by deemphasizing the quantitative aspects of science. More and more textbooks and courses have appeared that purport to teach science "qualitatively." This is deceptive advertising, albeit not intentionally. There is a big difference between learning science (quantitative) and learning about science (qualitative). We appreciate a great painting not just because it is beautiful, but because we ourselves as children have drawn pictures; we have "done" art, we know how much talent and training is required to be a creative artist. We need a similar experience with science; we need to "do" science in order to truly appreciate its power.

One of the most frightening consequences of science anxiety is the scientific illiteracy of the average citizen. We are all confident in our ability to make decisions about such issues as taxes, school bonds, and candidates for office. But what about building a nuclear power plant near our town? How dangerous is it really? Are the claims of environmentalists about the dangers of industrial pollution exaggerated or are they accurate? Will hair spray destroy the protective layer of ozone in our atmosphere? Is solar energy really safe? Increasingly, we are called on to make political decisions about technological issues, and we are not competent to do so -- not because we cannot understand science, but because we *believe* we cannot understand science. So we defer to the experts. And each side of a controversy produces its experts who bombard us with statistics until we tune the whole business out, thereby abdicating our power to affect decisions that affect our lives and the lives of future generations.

Can we really learn enough to make intelligent decisions about science-based issues? Can we sort out the claims of the experts and choose an appropriate course of action? Yes. You don't have to be a doctor to know when someone's selling you snake oil. You can learn enough science to make you a responsible citizen -- once you overcome your science anxiety. Science anxiety can be overcome. It is a learned pattern of behavior; it can therefore be unlearned. In order to do this, we must first see what science really is, not what our overexposure to TV, movies, and comic books has led us to believe it is.

[1]M. Green, "The Anti-humanist Humanists," *Chronicle of Higher Education* 14:32 -- 33, 1976.

[2]C.P. Snow, *The Two Cultures and the Scientific Revolution*, Cambridge, England: Cambridge University Press, 1959.

[3]See, for example, S. Tobias, *Overcoming Math Anxiety*, New York: W.W. Norton, 1978 and references therein.

Reprinted with permission from *Chemistry* (now *SciQuest*), pp. 6-9, October 1978.
Copyright 1978 by American Chemical Society.

2
Science As A Liberal Art

In Western society we tend to divide knowledge into two broad categories: the arts and the sciences. The former includes the fine arts (painting, music, theater) and literature and languages. The latter includes biology, chemistry, physics, astronomy, and (usually) mathematics. In the broad gray area between the two are what are known as the social sciences: psychology, sociology, political science, and economics. These categories reflect the recognition that there are differences in our approach to knowledge. One difference in these categories, for example, is the degree to which they lend themselves to quantification, with the sciences the most quantitative, the arts the most qualitative, and the social sciences somewhere in between. This, of course, is not a judgment; the arts *should* be qualitative, the sciences quantitative, and the social sciences somewhere in between. But one of the root causes of science anxiety is an overemphasis on the differences, and a lack of recognition of the similarities, leading to the extreme belief that there exists a "scientific mind" and an "artistic mind," and that the two are mutually exclusive. In what follows, both the similarities and the differences between the sciences and the arts are explored, and a number of widespread myths that frighten people away from studying science are debunked.

Differences Between Science and Art

Let us begin by listing the actual differences between science and art (in its broadest sense). First, science emphasizes quantity, art emphasizes quality. Science is always concerned with the answer to the question "How much?" We must recognize, however, that science has its qualitative aspect and art its quantitative. For example, the beginning of any science is the recognition of qualitative categories. Biology begins with taxonomy: the classification of living things by genus and species. Only after the qualitative distinctions have been understood can quantitative "laws of nature" be discovered that explain the classifications. The beginning of physics was Newton's recognition that there are two qualitatively different states of motion: accelerated and unaccelerated. Artists, on the other hand, are continually aware of a quantitative component in their field. How much of each pigment should be used? Is there too much light in the painting? What is the rhythmic structure of a poem? Nevertheless, we recognize that science is essentially quantitative, and art, qualitative.

Another difference is found in the tasks that the scientist and the artist set for themselves. The scientist seeks to understand and describe nature's order; the artist seeks to impose a personal order on the external world. Let us examine this difference more closely. The scientist, in order to be a scientist, ascribes to a number of articles of faith. One is that the universe is orderly, another is that we can comprehend this order. These beliefs are just that: beliefs. They cannot be proved either true or false. (Many scholars -- scientists and artists -- have considered what the consequences are of these beliefs.[1]) The artist has no such constraints. The world depicted by Picasso is every bit as valid as that depicted by Raphael. There need be no assumption of an external order; in fact, many modern painters and poets do not accept such an order at all. Thus the artist and the scientist are seeking entirely different kinds of "truths."

A third difference is the method for evaluating scientific versus artistic contributions. A work of art or literature is judged primarily on the basis of "taste." This concept is not very clear to those of us who are not artists. We believe that great artists, writers, and critics have "taste" and that they evaluate works of art on this basis. We also recognize that there may not be consensus on what constitutes good or great art; that there are often controversies and different schools among artists and critics themselves. Raphael's works are a standard of great Renaissance art, but other artistic styles are equally legitimate. "Right" and "wrong" have no meaning in art.

In science, on the other hand, there is a "right" and a "wrong." We believe that Nature is orderly, that she only does things in one way. As various scientific researchers try to find this one way, we must have a scheme for telling who has found the right way. This idea was brought home to me when I was an undergraduate physics student at Columbia, taking a seminar with Nobel laureate Polykarp Kusch. His opening remarks were, as I recall, "I don't really know what a physics seminar is. In a literature seminar, Professor T--- has an idea, and you have an idea, and you both discuss your different ideas. But in physics, either you're right or you're wrong!"

The evaluative scheme that scientists use is called the "scientific method." Most textbooks describe this method as the way scientists do science. It is not, as will be explained shortly. It is the way that scientists evaluate scientific work. In crude terms, it provides the ground rules within which the game of science is played. It does not provide a prescription for carrying out scientific research.

The scientific method contains the following components:

1. Observation of natural phenomena.
2. Development of several hypotheses that might explain the phenomena.
3. Performance of experiments to decide which of the hypotheses is correct.
4. Elaboration of the correct hypothesis and formalization into a theory that
   a. explains all hitherto observed phenomena and experiments.
   b. predicts the outcome of new experiments.
5. Carrying out of the new experiments to test the theory.

For example, let us develop a law of falling bodies according to this scheme:

1. We observe that many objects fall.
2. We hypothesize either that all objects fall at the same rate or that they all fall at different rates; for example, heavier objects fall faster.
3. We experiment with dropping stones of different weights from the same height. We find that they all hit the ground at the same time.
4. We elaborate the correct hypothesis into a theory: "All bodies fall to the earth at the same rate, independent of their "weight." We predict that future experiments will bear this out.
5. We drop more objects. Most obey our theory. Some, such as paper or feathers, drop more slowly. We therefore modify our theory to include effects of air resistance: "In the absence of air resistance (i.e., in a vacuum), all bodies fall to the earth at the same rate."

At this point, the whole business probably appears very simple. If that's all there is to it, why, anyone can do science! However, there may be some questions lingering at the back of your mind, such as: How did we know to pick only two possible hypotheses? Are there others? Why is the addition of air resistance considered simply a modification of the basic theory, which only holds in a vacuum? After all, our observations are not made in a vacuum but in the presence of air. And why is the question of how bodies fall an important one? It is in these sorts of questions that the creativity and artistry of science lie. Choosing the right questions and the right approaches, theoretical and experimental, is where the concept of "taste" enters science. More will be said about this a little later. Right now, the important point to realize is that the scientific method does not provide the answers. However, when a scientist has discovered a law of nature, he or she publishes this discovery. The format in which the discovery is described conforms to the scientific method. The author states the problem, outlines the possible hypotheses, and then describes the method by which the problem is solved; that is, the correct hypothesis is isolated. Often, but not always, the author predicts the outcome of future experiments based on the new theory.

The essence of scientific experimentation is that it must be reproducible; other researchers repeating the same experiments must get the same results for the theory to be valid.

The process just described is rarely carried out by one scientist or even one group of scientists at one time. It is more usual for one scientist to suggest what an important problem to study might be. Others may then perform some experiments. Yet others may propose several theories to account for the observed results, while still others do more experiments to distinguish between competing theories. When the correct theory for explaining all the results finally emerges, a different set of researchers may predict results of future experiments and a new group of experimenters carry out the suggested research. Each step in this process can take months to years. But at the end of it, we have a clear notion of "right" and "wrong."

A fourth difference between science and the arts is in a very concrete way a catalyst for science anxiety. It is this: Science learning skills are different from the skills needed to learn the arts. This simple statement is almost never articulated to students who are beginning to study science. Science is suddenly presented as a new subject. Students do not realize that the old skills that they have developed for the arts do not work for science. Later chapters discuss in detail the skills needed for learning science. Let us at this point simply enumerate a few. First, the sciences in general, and the physical sciences -- chemistry and physics in particular -- rely relatively little on memory and a great deal on logic. The student who attempts to memorize every "formula" in a physics course will fare much worse than the student who learns a few basic concepts and then concentrates on applying them to various physical problems. (In fact, the word "formula" is itself misleading: It implies a recipe or prescription; if only you know it, you can solve science problems. Not so. Science is not simply a series of cookbook recipes. In fact, as any creative chef knows, neither is cooking.)

A second "skills" difference between science and the arts is the way in which they are learned. If, for example, you are studying a Shakespeare play, your first step is probably to read through the play fairly quickly, just to get the gist of it. Later you may go back and look beyond the simple story line to the use of language, subtlety of metaphors, and multiple meanings of various passages. After numerous rereadings, each of which gives you further insight into the meaning of the play, you may be said to "understand" it.

Science is studied very differently. Although the total amount of

---

time you spend understanding -- fully understanding -- a scientific concept may be the same as the total amount of time spent in understanding the Shakespeare play, by far the largest chunk of your time in science comes during the first reading. You cannot skim a page of science textbook and expect to get much out of it. You must read it very slowly, pencil and paper in hand, and try to digest each concept as it comes. One page may take an hour. Each page is chockfull of new concepts. Professional scientists do not speed-read science: They read slowly and carefully, making notes and testing concepts as they go. They are accustomed to the slowness of the process. Students new to science, on the other hand, expect to read at the same rate they read literature. They quickly blame their own lack of intelligence when they cannot do so. The science-anxious student is probably thinking even while reading the science book, "What's the matter with me? Why is it taking me so long? I'll never get it." The answer is, it's supposed to take that long.

A third distinction between science skills and art skills may be stated as follows: In the arts, you know what you know; in the sciences, you may not. Let us illustrate this with our example of the Shakespeare play. Imagine you have read the play through once or twice. You know that you know the plot and that you have some familiarity with the use of language. You know that you do not know all the subtle metaphors or the archaic Elizabethan references. But you are comfortable in both your knowledge and your lack of knowledge of the play. You are probably confident that further study will reveal as much about the play as you wish to know. You feel grounded, balanced, secure in your ability to grasp Shakespeare.

Now consider a similar situation in studying, for example, chemistry. You may read a chapter of a chemistry text slowly and carefully, understand the concepts as they are presented, even understand the sample problems that the author works out as applications of the concepts. Now you attempt to work out some of the problems yourself -- but you cannot do so. Most of us have at one time or another encountered this situation, and we recall that panicky feeling when we just couldn't get it. We don't even know what we don't know. What we are up against is the reality that in a way science is like sports: There's a world of difference between reading about tennis, or watching tennis, and actually playing tennis. You learn science by doing science. There is an interplay between the text and the problems that is characteristic of science. The successful science student moves back and forth between the two until the concepts are truly mastered and applied to problem solutions.

There is a range of skills necessary for learning the various sciences. Biology depends somewhat more on memorization than do chemistry and physics. (Still, the sciences are more like each other than any one of them is like the arts.) People who are science anxious tend to be so in one but not all the sciences. Biology-anxious people may be quite relaxed about learning chemistry, and vice versa.

There are skills peculiar to the sciences. How do you get the most out of a science lecture? What if you miss an important point? Should you

---

focus on it at the risk of missing subsequent points? How do you go about solving science problems? What skills are necessary to perform laboratory experiments sucessfully? How can you get the gist of a newspaper article about science even if you don't understand all the details? All these questions will be dealt with later. Right now we simply enumerate them as examples of the actual differences in skills between science and art: differences that lead *unnecessarily* to fear of science.

Myths About These Differences

Having just discussed the actual differences between the arts and the sciences, let us now explore the myths about science to which these differences give rise. One myth is that science is an uncreative process, a kind of machine into which you grind some experimental data and out comes a theory. The nature of this machine is not clear. It probably has to do with "formulas." Science is a list of formulas; if you can only learn them, then you can do science. Perhaps the machine is the "scientific method." Scientists are people who have learned the steps of the scientific method; they apply it to their data, and presto! produce a new theory. These and other myths abound. Scientists are seen as people who have perverted their natural instincts into an emotionless study of objects. Or, yet another myth: There are such things as a scientific mind and an artistic mind, and the two are mutually exclusive. You are born with one or the other; no amount of labor will permit you to succeed if you don't have the right kind of mind.

One final myth: Science is "rigid" while art is "free." That is, there is no room for creativity, flexibility, or uniqueness in science the way there is in art. Science is just a set of rules. People with scientific minds can learn the rules, others cannot. Great artists each create something unique, while scientists are more like interchangeable parts.

It is not hard to see how these myths can arise from the actual differences in art and science. When we first study literature, we are given assignments to write essays about the books we are reading. Each student's essay is a unique creation. No two are alike; different ideas are equally valid. When we first study science, we are given problems to solve. There is only one right answer. There may be several ways to arrive at this answer, but even this fact is not emphasized in basic science courses. It usually looks to students as if there is one answer, one way to get it, and the person with the "scientific mind" is the one who sees that one way. (This is called "convergent thinking"; we shall return to it later.) Art and science at large, not only in the classroom, actually demonstrate this distinction. Monet painted unique works of art; he didn't simply paint the right picture first. Einstein, on the other hand, discovered the special theory of relativity first; if he hadn't, then sooner or later someone else would have found this "right answer."

This actual difference between art and science, the distinction between the artist's individualistic view of the world and the scientist's search for nature's order, gives rise to the myth that scientists are interchangeable, like machine parts. Even the social scientists, who attempt to employ the powerful research tools of the scientists, seem to view their work as more human, and more unique, because it deals with people rather than with things. In fact, the real distinction between the social sciences and the hard sciences is the rather inconstant behavior of people compared to the constancy of atoms and molecules. And who is to judge who is the more human? The biochemist who specializes in cancer research or the psychologist who develops techniques for manipulating people? Or the artist such as Leni Riefenstahl, who used her prodigious film-making talent to glorify the Third Reich?

There is no way to separate intellectual and artistic endeavors on the basis of their humanity. One can, however, judge individual scientists and artists by the moral application of their works. And here we must clear up another misconception about science. Science is not technology. Technology makes use of some of the discoveries of science. If technologists misuse those discoveries, then they must be faulted, not the scientists. The original scientific discovery is itself usually value-free; the uses to which it is put are not. Let us take for example the discovery of nuclear reactions: fission and fusion. On the one hand, it led to atom and hydrogen bombs, which we may agree in principle are bad. On the other hand, it led to nuclear reactors, which we are not sure are good or bad. But then again, it led to further understanding about the internal workings of the sun and other stars, to knowledge about the origins of the universe, to treatment of tumors by radiation. Science, like sociology, psychology, or art, can be employed for moral, immoral, or amoral purposes.

Debunking The Myths: Similarities Between Science and Art

We have explored some of the differences between science and art and the myths arising from these differences. We can now begin to debunk these myths by considering the similarities between science and art. We shall see that the title of this chapter -- "Science As A Liberal Art" -- is no misnomer.

What are the hallmarks of artistic creativity? Intuition, taste, originality, certainly. An ability to see things differently from others, to see new connections between old ideas. A sense of artistic balance. Courage to break new ground. And, somehow, an ability to reach people, to affect the way they think and feel. All these qualities are shared by the creative artist and the creative scientist. Let us illustrate this with a series of examples.

Consider first the notion of originality. As we said earlier, Monet painted a unique picture, while Einstein was merely the first to discover Nature's laws of special relativity. It would appear then that the artist Monet is more original than the scientist Einstein. Let us, however, look a little more closely. We may start by considering the paintings of Edouard Manet, an older contemporary of Monet. Manet painted in several styles. But one of them is certainly an early form of Impressionism, the style that

Monet so brilliantly brought to fruition. Monet, therefore, built on the past, adding his unique interpretation. This is a form of artistic creativity. Einstein did the same. He knew the laws of classical mechanics and electromagnetism as propounded by Isaac Newton and James Clerk Maxwell. He may have known there were certain experimental results, very few to be sure, that seemed to contradict these hitherto well-established theories. Many other scientists worked diligently to reconcile the new experiments with the old theories. But it was Einstein who made the creative leap. He saw that there was a need for the rejection of traditional notions of space and time and that a new theory would have to be crafted, one that not only explained new experiments but that would reduce to the old theories of Newton and Maxwell in the domain in which they were applicable. This domain, Einstein realized, was the region of fairly low velocities, velocities much less than the speed of light. Newton's laws of mechanics worked perfectly well for objects moving at low velocities. Thus Einstein's new theory, the special theory of relativity, explained the new experiments, which dealt with the motion of light, and reduced to Newton's laws for slowly moving objects.

Is this really different from Monet's rejecting earlier artistic styles while at the same time building on and extending the ideas of other artists? And just as later artists would express their creative impulses in the extension of Monet's Impressionism, so would other scientists extend the special theory of relativity to account for more and more phenomena. (It is, by the way, a myth that only a handful of scientists understand Einstein's theories. Each year thousands of research papers based on his theories are published.)

But isn't there a higher level of originality in painting a unique work of art than in simply being the one who got there first to discover a law of nature? To answer this we turn to another aspect of scientific creativity: the unique way in which each scientist elucidates a newly discovered law of nature. Nature's laws are not always obvious. We saw this earlier in our example of the falling bodies. Dropping a stone, a feather, and a piece of paper on earth would not lead us to the simple conclusion that bodies tend to fall at the same rate. It took the genius of Newton to recognize that the presence of air resistance was merely a necessary but minor modification of the fundamental law that in a vacuum bodies fall at the same rate. To take another example: Newton's First Law of Motion states that a body in motion tends to remain in motion at the same speed unless acted on by an outside force. This is again not obvious from everyday experience. If we slide a block of wood across a table, it does not remain in motion at the same speed; it slows down and stops. This slowing down and stopping was taken as the natural state of motion by Newton's predecessors, who thought that a body remained in motion only when acted on by an outside force; that is, if we keep pushing the block, it will move at a constant speed. Newton realized that the reason the block slowed down was that there was already a force present: the friction between block and table. Remove this (in the same way that you remove air resistance) and you get the actual law

of motion. It is this uncanny ability to strike at the heart of the problem that is the root of scientific originality.

How does the creative scientist decide what is the heart of a problem and what is simply detail? In fact, how does a scientist decide what problems are important to solve? It is here that we encounter intuition and taste, notions that are very similar for art and science. As we said earlier, scientists do not discover new laws by using the scientific method. The method simply provides a framework for science, just as canvas and oils and some basic mastery of technique provide a framework for creative art. Nor do scientists make discoveries by the use of mathematics, any more than poets write poems by the use of English. Mathematics and English are the basic tools of the trade; the creativity comes from within the creator. A creative scientist relies on intuition and taste to uncover Nature's secrets; he or she then communicates this discovery using the language of mathematics or the chemical or biological shorthands peculiar to those fields.

The scientist's idea of taste is remarkably close to the artist's. The artist seeks balance in a painting or a work of literature; the scientist seeks symmetry in nature. In fact, the scientist has a hard-to-justify belief that the laws of nature are simple and symmetric. If a theory to describe certain phenomena proves to be cumbersome, scientists will feel uneasy with it and will look for a simpler, more "elegant" explanation, to use their own terminology.

Einstein discovered the special theory of relativity because he was uncomfortable with a lack of simplicity in the Newtonian laws of motion and the Maxwellian laws of electromagnetism. This lack of simplicity, this asymmetry, is easy to describe. The laws of motion are independent of reference frame; the laws of electromagnetism seemed to depend on the existence of an absolute frame of reference. If, for example, you were on a smoothly moving train, moving at a constant speed with respect to the ground, there is no experiment involving motion that would allow you to tell if the train were in motion or standing still. (We assume that the shades are pulled so you cannot see the ground whipping by.) If you dropped a ball on the train, it would fall at your feet. If you weighed yourself, you would weigh the same as on the ground. This is what is meant by the "relativity of motion." Maxwells' theory of electromagnetism, on the other hand, predicts the existence of a fundamental speed: the speed of an electromagnetic wave, which we call light. This number is predicted unambiguously by the theory: 300 million meters per second. But if there is an absolute speed, our common sense tells us that it must be with respect to an absolute reference frame. Therefore, we would expect that experimenters on the train would, by measuring the speed of light, be able to distinguish between a train moving and one at rest.

Einstein was most uncomfortable with this lack of symmetry between motion of objects and motion of light. It offended his sense of taste. His intuition led him to postulate that the laws of electromagnetism, like those of mechanics, were independent of choice of reference frame. This led him to some remarkable conclusions: that the speed of light was measured to be the same number by the experimenter on the train and the one on the ground; that common sense notions of space and time were in fact not correct; and, finally, that mass and energy were one and the same. All experiments to date have proved him right.

Earlier we mentioned courage as a prerequisite for scientific and artistic creativity. The inner strength to break with past ideas is perhaps the most obvious characteristic of the artistic or scientific genius. Einstein's courage in defending his aesthetic judgment that the laws of classical physics were asymmetric is in some ways even more remarkable than his theory itself.

Aesthetic judgment, or taste, determines for the scientist which questions to ask, which avenues to explore. Perhaps the best analogy here is with chess. Computers have become more and more sophisticated; many can now be programmed to play chess. To date, however, any good chess master can beat any computer. Why? Because the computer is relying on something like the scientific method: It searches out a range of possible moves and calculates which is the most effective. The chess master relies on skill, but also on intuition: No broad searches of any alternatives, but rather a sense of taste coupled with years of experience seem to give the master the edge. Taste is what prompts the scientist to ask the right questions. Newton observed an apple fall. He also observed the moon orbiting the earth. His sense of taste led him to ask: "Are these manifestations of the same sort of motion?" His intuition and his mathematical skill led him to the law of gravity, which explains both the moon and the apple.

In some cases, scientific intuition may be so good that it leads to questions that no one had even thought of. If Einstein's special theory of relativity may be likened to Monet's break with realism and the development of Impressionism, then Einstein's general theory of relativity may be likened to the Cubist revolution in art, to Picasso's works, or to the writings of Gertrude Stein: so different from what had gone before as to be incomprehensible at first glance. Einstein considered the problem of reference frames accelerating with respect to each other, like a train picking up speed. He came to the remarkable conclusion that there is no way to tell if you are in an accelerating frame or merely in a frame at rest in a gravitational field. Acceleration and gravity are fundamentally indistinguishable. This notion may seem preposterous, yet it appears to be correct. Furthermore, the basic aspects of this theory are quite comprehensible to nonscientists and are sometimes taught in basic physics courses. Cubism is also at first glance somewhat preposterous, but the serious student of art will eventually come to appreciate the brilliance of the style.

The example of the general theory of relativity is instructive. Here is a case where the scientist did more than simply get there first. It was his unique creation from start to finish. And just like the Cubist painters, who had to develop new techniques, so Einstein had to go and learn new mathematical skills in order to work out the consequences of his remarkable intuition.

Einstein is, of course, not an isolated case of the scientist's search for symmetry and simplicity. Important branches of chemistry and physics deal entirely with the problem of symmetry in atoms, molecules, and elementary particles. Biologists look for the simplest mechanisms for explaining natural phenomena. One example of a symmetric and simple theory, one to which we have alluded a number of times, is Maxwell's theory of electromagnetic waves. In Figure 1 we see a reproduction of his equations, in a context that is both amusing and profound. Look at the simplicity of the equations. Even though you may not understand the symbols, you can immediately see that there are only four equations, that they look similar to one another, and that they are rather short. It is these aesthetic features that are so appealing to scientists. Not only do the equations explain all the measurements known, but they do it simply and elegantly. They lead naturally to the existence of waves of light. The caption, "And God said...and then there was light," expresses the belief of the scientist that Nature's laws are simple and beautiful, that the true miracle of light is the simplicity of the equations that describe it.

17

Figure 1. The "Maxwell's Equations" T-shirt. An example of aesthetics in science.

Science and art not only have the elements of aesthetic taste, intuition, and originality in common, but they progress historically in the same way: by cycles of innovation, development, and further revolutionary innovation. Each new revolution builds on some of the previous stage but rejects other parts of it. The Classical music of Haydn and Mozart both stem from the breaks with the Baroque style of Bach. Beethoven and Schubert bring the Classical period to its richest heights. Then comes the Romantic revolution, with Brahms, Liszt, and Chopin among its major innovators. This gives way to the Impressionism of Debussy and later to the atonal twentieth-century music of Schoenberg and Berg. These in turn are replaced by a Neo-Classical revolution, characterized by works of Stravinsky, Hindemith, and Copland. So it goes in science as well -- with, however, the additional proviso that revolutionary new theories must not only explain new experiments where the old theories have failed but must also agree with the old theories in the domain in which they succeeded. Einstein's special theory of relativity must get the same answers as Newton's laws of mechanics for slowly moving objects. (It does.)

It is also no coincidence that scientific and artistic revolutions seem to occur at the same times and to be quite similar in their objectives. The advent of quantum physics, with its emphasis on the probabilistic nature of things and on the interaction of experimenter with experiment, goes hand in hand with the development of Impressionist and Post-Impressionist art, with its emphasis on the artist's personal view of the world, as contrasted with the photograph-like painting of the early nineteenth century. Cubism appears in the Age of the Machine, the early twentieth century, with technology as its deity.

There is one more similarity between science and art: They affect the way people think and feel, often on a level below consciousness. In particular, changing styles in science both reflect and affect changing ways of life. Taking a crude example, martial music tends to arouse feelings of patriotism and love of country. This is its purpose. Great speeches can move millions of people to action, and not always in their best interests. In subtler ways artistic and scientific styles do much the same thing. The elegant piano pieces of Chopin were one with the central European salon life that they reflect, just as the folk motifs of the Russian composers represented and influenced society under the Czars. In a similar way, scientific discoveries affect the way we think and live. The most obvious example is nuclear energy, a result of the theories of relativity and quantum mechanics. The explosion of the atom bomb ushered in the Atomic Age. Gone was the optimism of nineteenth and early twentieth century civilization, with its faith that technological advance meant progress. A scientific revolution that began some three decades earlier with relativity and quantum mechanics now led to a revolution in our way of life.

Another example is Darwinian theory. Two ideas of this theory have had profound effects on civilization: the origin of humanity in "lower" species and the idea of natural selection. The former has led to theological debate that continues to this day. One might say that the acceptance of our origins in lower primates has injected humanity with a degree of humility: We are not so far from the apes, nor are we necessarily the apex of biological creation. Science fiction, and other literature as well, has treated the question of what the next step above us on the evolutionary ladder might be.

The principle of natural selection has had a somewhat more sordid history. It is an example of the way that a scientific idea may be taken out of context and used for pernicious purposes. Once Darwin had proposed "survival of the fittest" as the primary mechanism for evolution, this idea was taken over to justify racial discrimination and genocide. Learned articles purporting to show that the brains of women and nonwhites were smaller than and therefore inferior to those of white males were quite fashionable until rather recently. The murder of cripples, the mentally retarded, and "inferior races" such as Jews and Gypsies was justified by the Nazis on the basis of survival of the fittest and natural superiority of the so-called Aryan race.

These few examples show that scientific ideas tend to influence society, even if the members of the society have little or no knowledge of science. In our society today, people both admire and fear science. In subsequent chapters, we shall of course focus on the fear of science, on science anxiety. Before we do so, however, we need to understand the way, in social terms, this fear has become so widespread. We need to briefly explore our scientific heritage, to see how attitudes toward science have developed over the centuries and how the progress of science itself has both shaped and been shaped by these attitudes.

The Scientific Heritage

We begin with Aristotle, not because he was the first to question the form of Nature's laws, but rather because he represents the culmination of Greek "natural philosophy." Much influenced by the elegant, symmetric, and simple rules of geometry of the Pythagoreans and Euclideans, Greek natural philosophers sought similar rules for the workings of Nature. Thus the sense of aesthetics that modern scientists share became part of our heritage quite at the beginning of science. Unfortunately, the other crucial element of science, the subservience of theory to experiment, was not part of the Greek world view. Aristotle was supremely uninterested in experimentation. He sought only the simplest and most elegant theories, according to his aesthetic sense, and pronounced these Laws of Nature. He was not even especially interested in the geometrical views of his predecessors; he worked with logic alone. Thus his assertion that heavier bodies fall faster than light bodies is an assertion very easily tested and disproven *once one accepts experiment as the final arbiter of truth.*

Since Aristotelian philosophy had been accepted by the church, it took an act of courage for Galileo to actually make the test and proclaim that Aristotle was wrong, that if we neglect air resistance (and we can for two stones of different weight), all bodies fall at the same rate. We may credit Galileo with introducing the critical elements of experiment and observation into science. We may in fact trace the origin of what we call modern science to him.

It was also believed that the earth was the center of the universe and that all the other heavenly bodies revolved around it. The culmination of this belief was the Ptolemaic model for the solar system. By imagining concentric circles about the earth, other circles revolving around these, much like gears, and the sun and the moon and the planets moving on these various circles, Ptolemy could calculate the positions of all these celestial objects.

Nicolaus Copernicus proposed a much simpler model: that the sun was at the center and that the earth was one of the planets that orbited the sun. Each of these orbits, Copernicus believed, was a circle. And now we come once again to the question of aesthetic simplicity in science. For if we test the Copernican against the Ptolemaic model, the Ptolemaic model gives much better agreement with the observed positions of the sun, moon, and planets! Why then not reject the Copernican model? Because it is simpler. If only some minor modification could be made to bring it into agreement with experiment. That modification was made by Johannes Kepler, who substituted ellipses for circles. Further modifications, which included the effect of planets on each other, completed the process. (Galileo some decades later incurred the censure of the Inquisition for defending the Copernican model of the solar system.)

The rather sudden flourishing of science in the fifteenth to seventeenth centuries was a part of the European Renaissance, the emergence from the Dark Ages. It is again no coincidence that both the arts and the sciences began to flourish at the same time. When the atmosphere is ripe for creative ideas, they burst forth, with no regard for the particular "discipline" in which they occur. Furthermore, science and art were not viewed as separate, mutually exclusive entities, the way they are today. Science was a branch of philosophy on the one hand and a branch of technology on the other. The same Leonardo da Vinci who painted the Mona Lisa also designed a parachute, a helicopter, and a host of other devices. Kepler and Newton were not only interested in science but wrote voluminously on theology. Galileo built a telescope through which he viewed the moon. He reported his observations in the form of art, painting what he had seen in delicate watercolors.

In the eighteenth century science was still a product of Renaissance thought: as yet neither separate from the arts nor the special domain of an elite coterie of practitioners. In fact, to take a nice example, the founders of America were themselves interested in both the aesthetic and practical aspects of science. Benjamin Franklin is the best known example. Much of his success as Ambassador to France may be ascribed to his renown for contributions in the field of electricity. George Washington had technical training as a surveyor. David Rittenhouse, another of the Founding Fathers of America, was well known for his research in astronomy and optics.

"Look," I would say to Leonardo. "See how far our technology has taken us." Leonardo would answer, "You must explain to me how everything works." At that point, my fantasy ends.

Thomas Jefferson regarded law as his vocation and science as his avocation. His political philosophy is full of references to Natural Law and the orderliness of the universe, just as his Monticello home is full of the technological gadgetry of the eighteenth century. In the Declaration of Independence, he based part of the rights of the rebellious colonists on Natural Law; in this, he represented well the spirit of the times. Reason and science were rapidly replacing past authority and Divine right as the arbiters of an institution's legitimacy. If a government acted tyrannically, it was said to contravene the natural order of the universe, and this was a justification for its overthrow. Knowledge of science, like knowledge of art and literature, was considered an essential prerequisite for the person of culture.

With the striking successes of late eighteenth and nineteenth century science in the cure and prevention of disease, the development of new modes of locomotion such as the steamboat and the railroad (both based on the steam engine), and the fascinating and practical advances in chemistry, electricity, and magnetism, science began to take on the quality of a force for the salvation of humanity. The nineteenth century viewed science as a limitless source of treasure. Whatever problems beset humanity would be solved by scientific breakthroughs. The novels of Jules Verne represent very well this optimistic attitude toward science. Scientists gave popular lectures that attracted hundreds of listeners. There was no widespread notion that science was beyond anyone's comprehension.

In the late nineteenth century, science acquired an even more universal character. Until this time, only those who had independent sources of income could afford to engage in the practice of science. They had to purchase their own equipment and outfit their own laboratories. By the late nineteenth century, most European and a few American universities had realized the importance of science in their curricula. Bright young people from families of modest means saw the possibility of a career in science. The only requirement was a bright, inquisitive mind. Thus science became an equalizer, a democratizing force. Its practitioners in principle were judged solely on the basis of their accomplishments, not on their origins or class backgrounds. (In practice, racism, sexism, and anti-Semitism were not absent.) Most scientists grew up in families of modest means. The community of scientists today prides itself on its sense of fairness and democracy. This pride is somewhat naive. We are still beset with scarcity of women and certain minorities in the sciences. Nevertheless, it is better to have a democratic ideal toward which to strive rather than an aristocratic ideal that must be combatted in order to open a field to all who choose to practice it.

As the nineteenth century drew to a close, the methods of science were applied to new areas: the study of people and cultures, the so-called social sciences -- anthropology, psychology, sociology, economics, and political science. Here, as in the natural sciences, a rigid application of scientific method was not the path to creative advance. The good psychologist, for example, needs not only training but also intuition and insight into people, much as the good biologist or experimental physicist needs a flair for designing and building equipment. Naturally, one cannot and should not experiment on people in the same way as one does on atoms. As a consequence, results in the social sciences are not reproducible in the sense that they are in the natural sciences. Psychology, the most experimental of the social sciences, may reveal characteristics that vary from culture to culture. Anthropology and sociology are sciences of observation, like astronomy: One cannot do experiments on whole populations. And, as we mentioned earlier, the results of the social sciences may be perverted to evil ends in much the same way that the hard sciences may be applied to biochemical warfare and thermonuclear bombs. But these issues were all in the background at the end of the nineteenth century.

By the 1930's, two things had happened to science: It had made its most remarkable discoveries and it had become less accessible to nonscientists. The theories of evolution, of relativity, and of quantum physics had changed the world. Science had become more established as an intellectual enterprise worthy of lifetime study for those who chose it as a career--but more and more distant from those who did not. The era of the popular scientific lecture was over.

The model of the twentieth century scientist was Einstein. He was acknowledged by the whole world as a genius, yet people believed that almost no one could understand his theories. Einstein himself tried to combat this belief. He wrote, with Leopold Infeld, a wonderful book for nonscientists on physics from Newtonian mechanics to his own theories of relativity,[1] but he was unable to change the tide of events.

Other scientists were not so eager to share their new ideas with the public. Theirs had only recently become a respectable profession. Basking in their newfound (and well-deserved) prestige, they tended to devote their energies to making further discoveries rather than to popularizing those they had already made. Consciously or not, they jealously guarded their status by making science mysterious to nonscientists.

The public's view of science and scientists began to change. Coupled with admiration was a growing fear, at first a fear of the unknown and, with the explosion of the atom bomb, a fear of the known. Nineteenth-century views of the scientist as the savior of civilization gave way to twentieth-century fears of a runaway technology. The scientist became more and more a distant, mysterious, frightening character and less and less a role model to emulate. Science and technology became increasingly confused with each other, partly out of ignorance, but partly by the way that scientists "sold" science to acquire funding for their research. If they were to say, "What we are doing is similar to what artists and writers do," then they would be funded on the level of artists and writers, and research would come to a halt. So instead, they have said, in effect, "Look at our accomplishments. We gave you cures for disease and new products for your home. We gave you nuclear energy, the transistor, and the laser. Keep funding us, and we will give you more devices." The astronomer Simon Newcomb observed in 1886 that the government would fund science as long as it "received satisfactory assurance that scientific results are not the main object."[2]

Having once cleared up the difference between pure and applied science (i.e., technology) at the end of the nineteenth century, scientists have obscured that difference in order to get funds to do pure science. They and science are paying the price in public distrust and in general fear of science.

An interesting way to trace the various stages in the public's attitudes is to look at the depiction of science and scientists in films. We find almost from the beginning of movies a tension between the conflicting stereotypes of the scientist: savior of the world and cold-blooded monstermaker. We see science depicted at times as a liberating force, at times as the source of worldwide destruction.

An early talking picture, *Things to Come*, describes the redevelopment of civilization after some holocaust. The heroes are the technologists, the villains, the antiscientists. A good deal of the film, in fact, consists of speeches by Raymond Massey extolling the virtues of science and technology. A demurral from this point of view is presented in Charlie Chaplin's *Modern Times*, which deals with the alienation of humanity as a consequence of modern industrial technology.

A classic of the 1930's is the film of Mary Wollstonecraft Shelley's, *Frankenstein*. Boris Karloff's brilliant portrayal of the monster has eclipsed other features of the film; in particular, the degeneration of Dr. Frankenstein himself in the name of science. The doctor is the archetype (and the stereotype) of the scientist who starts out to save humanity but through his fanatical commitment to research winds up endangering the human race. In the original film the portrayal of Dr. Frankenstein is not unsympathetic. There is tragedy in his fall. Later versions, however, show the doctor in an increasingly unfavorable light. The rather balanced ambivalence of people toward science reflected in the earlier film has been replaced by a marked antipathy several decades later in the Atomic Age.

In the 1950's, we find a rash of "things from elsewhere" films: *It Came From Outer Space, Them,* a film version of H. G. Wells' *War of the Worlds*, and others. In these films, the scientist is generally portrayed as a hero, whose scientific brilliance rescues humanity from danger. (Gene Barry, the physicist hero of *War of the Worlds*, wears glasses to show he's a scientist, but takes them off to kiss the heroine.)

Other films, especially later ones, are not so favorable. In *The Thing*, we see the stereotype of the cold-blooded scientist, willing to risk the lives of everyone for a scientific discovery.

Although some film scientists save us from alien creatures, we are threatened by homegrown monsters created by others. Humanity is set upon by blobs, by giant crabs, and by teenage werewolves. When these are not the intentional products of some twisted scientist's research, then they are the unintended result of radioactive fallout.

The advent of the Atomic Age and the rash of public statements by scientists warning of the dangers of nuclear holocaust were reflected in a series of films that showed the scientist as a flawed genius, not like Frankenstein, whose obsession for research destroyed him, but rather like the actual scientists of the day, who, peacelovers all, helped produce the most terrible weapons ever known. They find representation in film versions of the works of Jules Verne. Captain Nemo, commander of the Nautilus, is one example. He appears in *Twenty Thousand Leagues Under the Sea* and again in *Mysterious Island.* In both films he is played by actors who are not necessarily typecast as villains: James Mason and Herbert Lom. *Master of the World* as portrayed by Vincent Price is another example. What do these characters have in common? They are scientists and masters of technology. They hate war and wish to end warfare. But their way of doing so is itself so violent that it eventually destroys them.

The ultimate antiscience film of the 1960's is *Dr. Strangelove.* All our fears of science are summed up in Peter Sellers' hilariously terrifying character: the bombmaker, the emotionless, maniacal genius, the Nazi experimenter.

The late 1960's and early 1970's brought us the "scientist as powermonger" and the "scientist as a willing tool of villains." The odd assortment of scientists who parade before us in, say, the James Bond films, provides a good example. From Dr. No to the bespectacled characters who are happy to produce any destructive device for money, we are treated to the most up-to-date stereotype of the cold-blooded, power-mad scientist.

But the most famous stereotype of our age, one who has taken on the dimensions of folk-hero, is the Science Officer of the *U.S.S. Enterprise*, Mr. Spock. Here is the distilled essence of scientific mythology: the infinitely competent, but emotionless thinking machine in (almost) human form. Much of the dramatic tension in *Star Trek* stems from the conflict between Dr. McCoy, the healer, the emotional, concerned human being, and Spock, the rational scientist. For an occasional variation, we see the conflict within Spock himself, between his human half and his scientific Vulcan half. Our response to Spock sums up our response to science in general: admiration mixed with no small component of fear and a great sense of relief on those occasions when the scientist succumbs and reveals his human side.

It is hard to predict what the film image of the scientist will be in the future. It would probably be a more benign image than Dr. Strangelove or Dr. Frankenstein, but certainly never again the white knight of the 1950's.

The Misuse of Science's Trappings

One of the spinoffs of twentieth century science, and one that increases people's confusion about science, is the flourishing of pseudoscience, such as astrology. Pseudoscience employs the jargon and trappings of science for its own purposes. It disregards the importance of experimentation, of scientific method, of reproducibility of results, substituting instead a spurious connection between all things in the universe. If the stars exist and we exist, so the notion goes, then the courses of the stars must affect the courses of our lives. The concept of "energy," which has a very clear meaning for chemists and physicists, is thrown about with reckless abandon by others. The science news section and the daily horoscope are treated with equal seriousness in the daily newspaper -- and more people read the horoscope.

Why does pseudoscience hold so much appeal for so many? First of all, it plays the same role as did primitive religion. In early stages of civilization, people had little control over the forces of Nature. When they had done all they could to avert natural disaster, they turned to the gods, entreating them to act benignly. With the development of civilization and the growth of modern science, this kind of direct appeal to objects gave way to a more sophisticated theology, one that deals with philosophical questions quite divorced from natural phenomena. From the Renaissance until the twentieth century, the apparently limitless advance of science and technology led people to think that they had a good deal of control over the forces of nature; they therefore had a decreasing need for the kind of primitive prayer just described.

In the latter part of this century it has become clear that not only do we not control Nature but that we can wreak more damage than any natural forces. There are three possible responses to this state of affairs: to study

science and attempt to bring technology under control, to avoid science and hope for the best, or to return to the primitive. This last choice is the choice of pseudoscience. If we can believe that mystical forces in the stars and the earth govern our lives, then we need not face the awful fact of our lack of power in the face of technology. The message that this lack of power can be remedied by the study of science falls on deaf ears, as long as people fear science, as long as they believe they cannot comprehend the laws of Nature. Those who simply avoid science without opting for pseudoscience are in the same bind; their tolerance for living on the edge is simply higher.

Why, if pseudoscience is antiscience, does it come clothed in the trappings of science? Because people know that science means power. The symbols themselves communicate a sense of power, even though the actual content is missing. The buzzwords of science have become so much a part of our culture that even those who fear and oppose science use them. Sounding scientific is still a way to achieve a kind of legitimacy.

Part of the impetus leading to fear and avoidance of science is its connection with what is called "alienation." Here is an area in which scientists and nonscientists seem to be completely at odds. Scientists do research to bring them closer to Nature, through understanding it better. But the public tends to equate science with separating people from Nature. This is in part a confusion of science and technology. Discoveries in biochemistry do not necessarily lead to an overdependence on drugs for all ailments; relativity and quantum mechanics do not mean that nuclear weapons must proliferate or that nuclear reactors need to be unsafe, but these distinctions are blurred in the public's mind. The role of computers in depersonalizing life and the role of technology in producing more sophisticated weaponry have certainly turned people away from science. Fear coupled with distaste for the perceived consequences of science have led to dropping enrollments in science courses. Not only is science hard, we are told, but it depersonalizes you.

And while we are once again on the topic of how difficult science supposedly is, we should realize that this myth is also an impetus for all manner of pseudoscience. To learn to do science is admittedly an arduous task (just like learning to do anything well). To memorize the buzzwords of science and parrot them back as a kind of dogma is not nearly so difficult. The astrologers do it; so do the purveyors of various political dogmas. To them, the theory is all; whether it agrees with observations is beside the point. In fact, the notion of an objective universe existing outside of us, the cornerstone of science, is the first victim of these theories. External authority and subjective feelings replace the drive for experimentation.

This discussion should not be taken as an attack on all nonrational experience. Certainly there is room for, and need for, experiences and feelings that transcend rationality. But these should not come disguised as science. Religion has confronted this issue head-on. It does not pretend to compete with science; it deals with a completely different range of experience. And there are many religious scientists. There are very few if any scientists who are astrologers.

Sometimes even serious students of science can lose sight of the essence when they are overwhelmed by trappings. There is only so much that statistics, for example, can tell us, and the calculation of these statistics using large computers does not necessarily lend them any more validity. Reams of computer paper are no substitute for the intuition and experience of the natural or social scientist. That is not to say that science has become less valid. But there is a danger that students of the natural and social sciences are becoming less aware of where the quantitative ends and the qualitative begins.

The Thrill of Science

It is appropriate to close this chapter by describing a mystical, ecstatic feeling that is peculiar to scientists: the feeling you get when you make a discovery, when you reveal a new secret of Nature. Many great scientists have described this experience. To quote only two:[3] Here is Werner Heisenberg talking about his discovery of the Uncertainty Principle and the matrix formulation of quantum mechanics:

> I became rather excited, and I began to make count-
> less arithmetical errors. As a result, it was almost
> 3 o'clock in the morning before the final result of
> my computations lay before me...

> At first, I was deeply alarmed. I had the feeling that,
> through the surface of atomic phenomena, I was look-
> ing at a strangely beautiful interior, and felt almost
> giddy at the thought that I now had to probe this wealth
> of mathematical structure Nature had so generously
> spread out before me.

And here is Albert Einstein, describing his discovery of the general theory of relativity:

> Scarcely anyone who fully understands this theory can
> escape from its magic.

[1]A. Einstein and L. Infeld, *The Evolution of Physics: The Growth of Ideas from Early Concepts to Relativity and Quanta,* New York: Simon & Schuster, 1938.

[2]Quoted in D.J. Kevles, *The Physicists: The History of a Scientific Community in Modern America,* New York: Vintage Press, p. 59, 1979.

[3]Quoted in S. Chandrasekhar, "Beauty and the Quest for Beauty in Science," *Physics Today* 32:26 -- 27, July 1979.

3
Science Teaching And Science Learning:
The Present State Of Things

Why Students Fear Science

In order to understand why so many people fear and avoid science, we must explore what their experience with science has been. The last chapter discussed the influence that society's stereotypes have on attitudes. This chapter explores the role of schools and teachers in shaping students' attitudes toward the study of science. We will look at science as it is taught in universities as well as in high schools and in lower grades.

Talk to science teachers at colleges and universities and you hear a common theme: Science enrollments are low; most students who take science only do it to fulfill some requirements; many students who do very well in other courses perform poorly in science. Although science departments at universities are no smaller than liberal arts departments, they have a more difficult time attracting students to their courses. The word is out: "Science is too hard." If students haven't heard this before they get to college, they certainly hear it in the first few weeks from their classmates. Although many colleges have a science requirement -- a certain number of credit hours -- students, except those majoring in science, will generally seek out the courses they have heard are the easiest. Certain science courses become known as the easy ones and are filled with humanities majors. Rare indeed is the English major who signs up for physics, much rarer than the physics major who signs up for literature courses.

Even among science majors, fear and avoidance of science are widespread. Many students, for example, major in biology in the expectation of going to medical school. These students are often not enamored of science and are under a good deal of pressure to get good grades; they therefore opt for the easiest science courses they can find. They are as fearful as their friends in the humanities, but they have less opportunity to avoid science courses. Their anxiety comes out in other ways: "clutching" on exams, inability to concentrate on the material of science courses, and even physical illness.

In general, the university environment is not very different in its science attitudes from society at large. It is faddish to be antiscience, to believe that it is too hard for most people to grasp and that those who do study science are odd or different from the rest of the world. It is not socially acceptable to flunk literature or history, but a failing grade in chemistry or physics is sometimes worn as a badge of honor. And it is not only students who believe and spread the stereotypes about science. Teachers of humanities were, usually not too long ago, students themselves. They have often kept the attitudes they had toward science in their student days. If they view science as an uncreative endeavor, and scientists as a breed apart,, can one expect their students to see things differently? Colleges are divided into humanities and sciences -- in different buildings and rarely interacting with each other. The professors themselves reinforce the stereotype of the "artistic mind" and the "scientific mind" as mutually exclusive.

There have been a number of studies on attitudes of students toward scientists. All these studies show that students view the scientist as a distant figure and, generally, not one they wish to emulate. In a study entitled "Changing Attitudes Towards Science Among Adolescents,"[1] students who were studying high school physics were asked to rank physicists and other professionals on the basis of a series of personal criteria. They rated physicists high in "maturity" and "importance" but very low in "friendliness." Another study, entitled "The College-Student Image of the Scientist,"[2] sampled student opinions at various types of colleges. The stereotype of the scientist that emerged was of a person who is highly intelligent but socially withdrawn, someone devoted to research at the expense of family, culture, and social life in general; a narrow practitioner of a powerful craft. And in a very surprising study of attitudes of arts and science students toward scientists, "The Stereotypical Scientist,"[3] *both* the arts *and* the science students agreed that scientists are highly intelligent but personally dull human beings! What this means is that some people choose careers in science in spite of, or perhaps even because of a belief that scientists are socially inept. Scientists themselves therefore fall victim to their own stereotype. The practice of science becomes an excuse for escaping from the everyday world. There is little doubt that scientists form their own social cliques and tend to share personal mannerisms in the way that ethnic groups do.

This is not to say that stereotyping is all there is to the story. There have been numerous studies on personality traits of people in various occupations. One study of scientists[4] shows that they have a predominance of certain traits such as (not surprisingly) curiosity, a desire for personal

Reprinted with permission from *Chemistry* (now *SciQuest*), pp. 6-9, October 1978.
Copyright 1978 by American Chemical Society.

independence and for mastery of the environment, an interest in how things
work, some aversion to personal relations with large numbers of people,
and an intense drive in their chosen profession.   Although these
characteristics certainly show that scientific stereotypes are not entirely
without basis, we must realize that scientists share these traits in varying
degrees, that other professions also share certain traits that are on the
average different from "ordinary" people, and most importantly, that *this
does not mean that people who do not fit the "scientist-personality profile"
are incapable of learning and enjoying science*. Furthermore, as the study
cited shows, other factors in a scientist's childhood affect his/her choice of
profession: love of learning for its own sake as a family ideal, experience
with gadgets while growing up, the ready availability of books, and so
forth. It is obvious that these conditions are in large part influenced by
cultural and societal conditions. Should these change, we may expect to see
scientists arising from new sources. In particular, the numbers of women
and disadvantaged minorities in the scientific professions will increase.

Although scientists may to some degree fit the stereotypes about
them, this should not prevent others from learning science -- as long as
neither group, scientists or nonscientists, becomes invested in over-

emphasizing the stereotypes. Unfortunately, as we have seen, this does occur in the media and other places. Later in the chapter, we shall explore the interaction of science teachers with their students and return to this issue.

Science might fare better at universities if those institutions themselves did not play a dual role in society. On the one hand their nominal task is to educate. If this were their only task, then students might be willing to risk taking a "hard" course in science rather than an easier one. But the university's other function is to certify: to award diplomas, to confirm that students are capable of studying medicine, dentistry, law, or business. With this sword over their heads, students are much less willing to take risks. The university itself therefore implicitly encourages them to opt for easier courses. Since science courses are considered hard, they are the ones to suffer.

What about science in elementary and secondary schools? College freshmen seem for the most part already to have science anxiety by the time they arrive at the university. What has happened in the lower grades to cause this?

If we were to ask what one word summarizes the skill that is most emphasized in grammar school and high school, the answer would be "memorization." We can all remember our school experience as a mass of facts to be learned: dates in history, words and grammatical rules in English, multiplication tables in math. Although to a great extent the ability to memorize is important -- one needs a background of facts before one can deal with abstract ideas -- memorization has been stressed to the detriment of other skills, in particular, skills needed for science. Science consists of marshaling a series of observations about nature, analyzing or breaking down the information in these observations to look for underlying similarities, and synthesizing or building up from these similarities a theory about how nature works. But the skills of analysis and synthesis are not taught. Science courses in the lower grades are generally descriptive. Emphasis is on laboratory demonstrations and often on the "gee-whiz" kind of demonstrations that keep students interested without teaching them very much. This approach gives students an erroneous view of science. Biology becomes a list of organs to be memorized, chemistry a list of reactions, and physics a list of equations. The true nature of science as a kind of game or puzzle to be solved is not made clear. Students can get by in science courses by memorizing the right things for the exams. Why should this be so? I think because many, if not most, of the science teachers in the lower grades believe the same myth as the rest of society: that the talent necessary for doing science is given only to a select few. Note that the same approach is never taken with, say, art or literature. Grammar school children are not all expected to become professional artists or writers; nevertheless, they are required to learn the rudiments of painting and English composition.

So memorization is rewarded in science courses in the lower grades. What else is rewarded? Something that psychologists call "convergent thinking," or "linear thinking." This is the kind of thinking that yields success on standardized mathematics and science tests: There is only one route from the question to the answer, and the convergent thinker finds that route. An example of a question that tests convergent thinking is "Detroit is 300 miles from Chicago. How long does it take a car traveling at 50 miles per hour to make the journey?" There is only one way to do this problem. Speed x travel time = distance traveled. So travel time is distance traveled/speed. This is 300/50 or 6 hours. Well, one may ask, isn't this the typical kind of word problem that constitutes math or science? Not really. Only in the simplest situations does one have three quantities of which two are given and the third bears an obvious relation to those two. The skill that distinguishes practitioners of science is a combination of the ability to think convergently and to think *divergently*, to marshal a series of apparently disparate observations and to see connections between them. Newton's realization that the apple and the moon were exhibiting the same kind of motion -- attraction to the earth by gravity -- is a beautiful example of divergent thinking. Unfortunately, what is usually tested on standard exams such as the College Boards, the American College Testing Program, and The Medical and Law School Admissions Tests is convergent thinking. Students are trained to pass these tests. Divergent thinking is entirely overlooked.[5]

We can quickly see the connection between science taught as convergent thinking and the stereotype of science as a machine into which data are fed and out of which comes a theory. They are one and the same notion. If there is only one way to approach a science problem, then a robot can do it as well as a person. In actuality, doing science is something like playing chess. Of the various possible moves, we must select one. This selection depends on intuition, on "taste," on experience (sometimes on luck), all very human qualities. The divergent thinker is one who can engage all these facilities in his or her approach to nature. However, divergent thinking is not easy to test for, since there is no "right answer." Convergent tests are scored on the basis of correct answers; divergent tests on such things as flexibility and originality. These obviously do not lend themselves to easy standardization. I must admit that I have no ready remedy for this dilemma. But the problem is clear. Whatever we are testing, it is not the ability to understand science. This is an area ripe for educational research. In the meantime, however, it is important for science teachers to recognize that preparing students for exams on convergent thinking is not the same thing as teaching them science. Chapter 5 discusses ways of actually teaching science, so that students get the flavor of the divergent thinking processes that make science creative.

Concrete Thought, Formal Thought, and Science Anxiety

Let us now turn to another problem of science teaching and learning in the lower grades (and to some extent in college as well). This is the problem of assessing when a student is intellectually mature enough to understand and to do science. An attempt at defining this problem was made some years ago by three physicists: Robert G. Fuller, Robert Karplus, and Anton E. Lawson.[6] They looked at the model of childhood learning developed by the great psychologist Jean Piaget and tried to determine at what stage of growth students were able to grasp scientific ideas (in this case, the ideas of physics).

Piaget's theory divides the intellectual development of the child into four stages. The first, which occurs during the first two years of a child's life, is the *sensory-motor* stage. In this stage, the child develops physical skills and begins to understand that material objects exist even when he or she cannot see them. The second stage, called *preoperational*, occurs during the next seven or eight years. In this stage, the child tries to put together various facts and construct explanations for events in his or her life. However, the child does not yet reason from cause to effect. The final two stages, the ones that are seen in young people above the age of 10, are the *concrete reasoning* stage and the *formal reasoning* stage.

Students in high school and college are usually in the process of passing from the concrete reasoning stage to the formal reasoning stage. This introduces a serious difficulty in the teaching of science.

To really do science, a student must be capable of formal reasoning; yet, many students have not yet entered the formal reasoning stage. Therefore, science as taught in upper elementary school and high school is generally taught at a level that can be comprehended by concrete reasoning. Science as ideally taught in college requires formal reasoning skills. The transition from one to the other is often difficult, even for the student who has done well in high school science. Thus, if a student's high school science courses have emphasized concrete reasoning, the student is shocked to find how different things are in college. I wish I had a nickel for every student who has come to me in one of my physics courses and told me "I had high school physics, but it wasn't *anything* like this!"

We can illustrate the difference between concrete and formal reasoning by considering an example. Suppose you were given a simple pendulum in the form of a ball hanging from a string, as shown in Figure 2. You might then be given the following instructions: "This is a pendulum. Find out which variables affect the "period," i.e., the length of time of the swing. Some possibilities are the weight of the ball, the length of the string, the height through which the ball drops, and the strength with which you initially push the ball."

The formal operational thinker will solve the problem by systematically testing the effect of all the variables, one by one, holding all the others constant, and will be able to show that only the length of the string has any effect on the period of the swing. The concrete operational thinker may be able to solve part of the problem; for example, may recognize that the length of string has some effect, but will be unable to systematically eliminate the other variables. The formal thinker, in other words, can take concrete observations and reason abstractly to gain insight into the principles underlying these operations. The formal thinker is

Figure 2. Pendulum used to test for concrete and formal operational thinking. The weight of the ball, the length of the string, and the range of the swing may be varied to see which of them affects the period of the pendulum's motion.

comfortable with the notion "all other things being equal...."[7] He or she approaches a problem such as that of the pendulum by making "all other things equal," except for the variable under consideration. First, such a student holds the weight of the ball, the length of the string, and the strength of the push constant and varies only the height from which the ball is pushed. Then, having concluded this set of tests, the formal thinker holds everything but the weight constant and varies the weight by placing various balls on the pendulum and observing if there is any effect on the period of the swing. And so forth.

The formal thinker can also think in terms of what is called "propositional logic." That is, he or she regularly analyzes observations in terms of "if...then...therefore." "If I vary the weight of the ball, then the period of the pendulum swing is unaffected; therefore, the period is independent of weight." "If I lengthen the string, then the period of the swing gets longer; therefore the period depends directly on the length of the string." Note that unless also required to make quantitative measurements, even the formal thinker will not be able to deduce the exact nature of this dependence. (The period of the swing varies as the square root of the string's length.)

Formal thinking is not restricted to science. It is required in any field that makes use of logical propositions, such as law. To quote Piaget himself, the formal thinker is "an individual who thinks beyond the present and forms theories about everything, delighting especially in consideration of that which is not."[8] From this brief quote, we can see that Piaget thought there is a natural tendency for people to develop formal thinking techniques; otherwise, they would not "delight" in doing so. If, therefore, the process of moving from concrete to formal thought is a natural one, we must ask two questions: "What is the range of ages in which students pass from one to the other?" and "Can science teaching aid or obstruct this natural transition?" Piaget claimed that the transition occurred between the ages of 11 and 15. This means that a majority of 15-year-olds should be capable of formal operational thinking.

However, in a series of fascinating studies carried out by Renner and coworkers, some very disturbing discoveries were made about science learning and thought processes.[9] First, they were able to show that the concrete-to-formal transition was a natural one. This they did by simply testing students in various grades without introducing any special courses in science (or in anything else). Then they surveyed the effects of various types of science courses on the speed of the transition from concrete to formal and concluded that certain types of science courses speeded up the transition while others appeared to hinder it. The courses that were the most useful were the ones that recognized that the student must be approached at his or her own level of development. There is no way to skip over concrete thought and simply become formal operational. And there is no way to develop concrete thought without concrete "hands-on" science experiences: demonstrations and laboratories. Furthermore, if most high school sophomores are concrete operational (and they are), then they must be taught concrete science. Formal abstractions, "unifying underlying principles," although appealing to the teacher, will be beyond them. Juniors and seniors are formal operational in larger numbers; therefore it is possible to introduce formal thinking in science courses for them.

Unfortunately, there are two ways to hinder the transition from concrete to formal thought. The first is to introduce formal thinking too early, before students have had sufficient opportunity to become familiar with concrete experience. This was an error made by some of the novel high school science curricula in the late 1950's and early 1960's. In their desire to teach "real science," they did not recognize that students were not ready, because real science is formal operational. The other way to hinder the transition is to continue to teach concrete science when students are ready for formal abstractions. This is the present situation in a great many high schools. The students who come to college and are overwhelmed by science courses are obviously not yet formal operational. Is this because their natural transition rate has been impeded by high school science courses? It appears that the answer is often yes.

Renner and coworkers were able to assess what proportion of students in various grades were formal or concrete thinkers. They found that about 73 percent of 10th graders are concrete operational, about 50 percent of college freshmen are still concrete operational, and about 40 percent of all Americans never become formal thinkers! This is pretty sobering information. If we look at how various subjects are taught in high school and college, we see that they are not at all geared to the level of the students. For example, 10th graders study geometry and learn to diagram sentences in English. These two operations are as formal as you can get, yet only a quarter of the students can think formally. Foreign language courses in high school are also often taught formally: grammar, syntax, and so forth. If they were taught concretely, with less attention to formal structure, they might very well be learned better.

What about science courses? In high school, they are generally taught at the concrete level; in college -- at the formal level. This presents a dilemma, since about half of the students taking college science are still concrete thinkers, while almost half of the high school seniors taking science are making the transition to formal thinking. Thus the needs of about half of each group are met neither in high school nor in college. The formal thinkers in high school are held back; in particular, they are rewarded by good grades on tests that are concrete. Thus even students who might naturally make the transition to formal thinking are not motivated to do so, since they have learned that the "right way" is the concrete way. In fact, standard college qualifying tests that test concrete, convergent thinking do not correlate well with formal operational thought as measured by Piaget-type tests. We should thus not be surprised that students enter college unprepared for formal courses in the sciences.

What happens to students when they enter college? As many as can avoid science courses. Of those that do not, how do the concrete thinkers fare compared to the formal thinkers? Does the traditional formal course help concrete thinkers make the transition to formal thought? Can a concrete thinker get through a formal science course?

The answers to these last two questions are no and yes respectively: just the opposite of what we might hope. Researchers[10] have shown that, for example, traditional formal physics courses in college do not help concrete thinkers become formal. The concrete thinker who enters a college science course often leaves that course still a concrete thinker. However, and this is the answer to the second question, the concrete thinker can still get through the course with a passing grade! Studies indicate that college science course grades do not correlate well with formal reasoning processes.[11,12] How can this be? If science is based on formal thought, and college science courses are taught formally, how do concrete thinkers get through? Well, one thing they do is suffer a lot. Instead of mastering concepts, they fall back on memorizing formulas. They become "technically literate"; they learn the buzzwords of science without ever learning how science is actually done.[13] Since it is not always easy for a science teacher to distinguish between learning a technical vocabulary and really learning science, these students can get by. Even more ominously, science teachers tend to grade exams "on a curve"; that is, they assume that the average is about C+ or B- and grade the class accordingly. Thus if enough concrete thinkers do poorly on a test, the class average drops and a rather low score becomes a B-. This kind of inflated grading procedure does not measure whether students have mastered science, except inasmuch as the students who obtain very high grades are probably formal thinkers. They can truly grasp such ideas as infinitesimals, limits, or ideal states of a gas; the others just know the right words.

So we see that science teaching can be a primary cause of poor science learning. No doubt it contributes to the anxiety surrounding the study of science. In fact, one study[14] found that only 25 percent of *in-service* science teachers were formal thinkers. If the teachers are concrete thinkers, how can the students be otherwise?

We are thus left with some serious problems in teaching science to students at various levels. Later in the chapter, some ways that science teachers are approaching these problems are discussed. There is, however, one more question we must ask: How large a role does anxiety play in keeping students in a concrete operational mode of thought? Are those students who are afraid of science somehow keeping themselves from becoming formal operational? Are they staying with the comfortable, familiar thought processes because they fear moving on to the next level? For that matter, has anyone checked whether anxiety can affect how someone scores on a Piaget test? Can a student who is actually formal operational, under stress of a test, "drop back" to concrete operations? From what I have seen in the classroom, the answer seems to be yes.

Students who are able to work homework problems requiring formal reasoning are often unable to reason on an exam. They therefore fall back on searching their memories for formulas. In a pinch, they are so frightened of science that they don't trust their own proven ability to master it. In fact, I would not be surprised if the large number of students who test as concrete thinkers on Piaget tests is due to science anxiety on the part of the students. When confronted with the pendulum experiment discussed earlier, those who are afraid of science will probably panic and perform unsystematically, thus appearing as concrete thinkers, even if in less stressful situations they can reason formally. Here is a research project for a psychologist: to administer a Piaget test and an anxiety test simultaneously and see if there is some correlation between level of anxiety and performance on the Piaget test. If there is, then we must first deal with the anxiety. Only after this is done do we have any hope of getting an accurate measure of students' reasoning processes.

Before we leave the topic of concrete versus formal thought, let us consider one more interesting bit of information found by Renner et al.[15] In an attempt to test the general validity of Piaget tests, such as the pendulum test mentioned earlier, they looked for a group that was formal operational but not science oriented. They finally chose law students, on the assumption that to be a good lawyer you must use formal thought: "If...then...therefore" and so forth. Their assumption was correct. Some 80 percent of the law students scored as formal operational on the Piaget tests. Now this was a group of people already in law school, with clear career goals and no particular incentive to score well in science-type tests. In other words, a group that was not science anxious. These students scored well on the tests for formal thought.

Let us ask the hypothetical question, how would these law students have done four years earlier, as college students? With the pressure on to get a high grade point average in everything, probably not nearly so well. In fact, pre-law college students, like other nonscience majors, are notorious for avoiding challenging science courses and for taking the simplest courses to fulfill their college requirements. So what do we have here? A group that, on the basis of Piaget tests after college, is obviously capable of comprehending science. Indeed, it is from this group that many of our political leaders are drawn: leaders who will have to make science-based decisions. Yet this same group has managed to avoid as much contact with science as possible because it mistakenly believes that science is beyond its comprehension!

Women, Minorities, and Science Anxiety

Let us now turn to the question of tracking of certain groups away from science. It is obvious that women and disadvantaged minorities are terribly under-represented in the sciences. Is there something that society in general and the schools in particular do to cause this? We can get a couple of interesting clues by returning to some of the studies of Piaget reasoning processes. In a study of college students, D. F. Griffiths found that there was no difference in the percentages of white and minority students who were concrete thinkers (60 percent for both groups).[16] Thus it is certainly not a question of different levels of Piaget reasoning, even given that the minority students were the victims of, among other things, worse schools. In fact, "poorer schooling" is a hard to thing to determine in the area of science. Science anxiety and avoidance certainly seem to transcend racial and class barriers. Yet in the final analysis, there are far fewer minority scientists than their proportion in the population. Something else must be operating since black and Hispanic students get a clear message that science is not for them.

What about women? It is widely regarded as true that girls mature intellectually earlier than boys. High school girls, it is said, are one or more years ahead of their male counterparts. Yet when researchers[17] administered Piaget tests to high school students, they found that the boys were somewhat more formal operational than were the girls. Whence then the notion of girls' intellectual maturity? Renner and Stafford suggest that girls are socialized to be more compliant, less forward, less "disruptive" than are boys. The girls quickly fit into the concrete operational mold, since this is obviously what the teachers want. A formal thinker in a high school class that operates at the concrete level would probably, if he or she were outspoken, be considered a discipline problem.

Here then is the situation. Minorities and women see very few scientists as role models. Very early they get messages that science is not for them. They are steered overtly or unconsciously by their schools away from science. At best, those who are not steered away are rewarded for being concrete operational. Their expectation of being able to do science is therefore correspondingly lower than that of white males (and that is itself none too high), and this expectation is fulfilled. The tragedy is not only that there are few women and minority scientists but that these groups are filtered out of a wide range of professions that require technical expertise and formal thinking, such as medicine and law. And this discrimination perpetuates itself: The next generation is again faced with a lack of appropriate role models.

Let us consider how the message of who can or cannot do science is communicated to people. I can draw from experiences related to me by students in our Science Anxiety Clinic:

Example 1. A sophomore woman in one of our groups began college in a pre-nursing program. She later decided that she actually wanted to be a doctor. She quickly found that the psychological support she had received from her parents eroded. All her female cousins her age were getting married. This was what her parents hoped for her. They had been supportive of her studying for a career -- as long as it was a traditional "woman's career," nursing, with the chance of meeting eligible men -- doctors -- in her work. Once the woman decided to become a doctor, however, she broke all the unstated rules. She was entering a traditionally male preserve; she would have to be in school so long that the prospect of an early marriage was dim; and by competing with men, she would alienate them, so her pool of available marriage prospects would go down. How did this woman deal with these pressures? She was not ready to confront her parents directly. She had been brought up to believe the stereotypes that she was only now engaged in contradicting. So she became anxious. She became especially unable to function well in those courses that were most crucial to her acceptance to medical school -- her science courses. Thus she found her way to our Clinic. Her story is of course not unique and partially explains why two-thirds of the applicants to our Clinic are women.

Example 2. A young man who came to the continental United States from Puerto Rico as a teenager entered college hoping to become a doctor. He apparently received a great deal of encouragement from home. He was the first person in his family to attend college, let alone medical school. He himself felt some pressure as a prospective role model, an example for other Hispanics. In his last year of high school, in spite of his good grades, his guidance counselor advised him to avoid science and to abandon hopes of becoming a doctor, because, in the bizarre logic of the counselor, Hispanics have not been successful in science! The young man recognized the racism inherent in his advice, yet the counselor was a figure of authority. So every day in college the young man asked himself, "What if the counselor was right?" This self-defeating message, coupled with the difficult adjustment he had to college, led to a marked drop in his ability to perform in science classes, and he joined the Clinic.

Example 3. A black woman, also the first in her family to attend college, found she was anxious in almost all her science courses. She reported that she received a great deal of encouragement at home, not only from her parents but also from the parents of her friends, who always asked her how she was doing in college. Upon reflection, however, she realized that she was getting a very mixed message from her friends' parents. On the one hand, they sincerely admired her. On the other hand, they saw her in contrast to their own children, who were more the norm for their community. They had internalized the oppressive messages that they were not college material in general and not science-minded in particular. Our science student sensed that she was doing something "wrong," and that sooner or later she would get her comeuppance. She responded by becoming anxious.

What all these students are doing, and what their communities are doing, is buying the societal myth that some people are born to have power and others are not. In this case the power is the power of intellect, the power that comes from mastering the techniques of science and using them to unlock the secrets of the universe.

Interestingly, there is another side of the coin of cultural, racial, or sexual oppression as applied to the study of science. There are certain ethnic and national groups that produce scientists in greater proportion than their numbers in the general population. Jews, Indians, Chinese, and Japanese come immediately to mind. Because of various cultural norms about knowledge and education, members of these groups have in general been able to surmount the obstacles placed in their path by racial and religious bias. There is much to be learned by a systematic study of the cultural factors that come into play in overcoming such obstacles. However, sometimes such factors actually work against the student from one of these cultures. One student in our Clinic, from an Eastern culture, was having difficulty passing any of her science courses. She hoped to become a professional scientist with a Ph.D. in her chosen specialty. Yet she was unable to function in a science class. When she began to reveal her background in the Clinic, it became clear that everyone in her family was a successful scientist or a professional: men and women alike. The pressure on her to follow in their footsteps was overwhelming -- and she had indeed been overwhelmed. Science was so highly regarded in her family and community that she had never had the luxury of deciding whether she wanted to do anything else.

These few examples indicate the complexity of the cultural forces that play a role in science anxiety. Much work needs to be done in the social sciences to really map the effect of these forces. What we can do here is to recognize their existence, and their potency, so that we may develop individual and group strategies to overcome them.

The Science Teacher: Role Model or Anxiety Producer?

We have discussed in a general way the effect schools have on the study of science and on the fear of science. Let us now focus specifically on the science classroom and particularly on the role of the science teacher in fostering science anxiety. The first question to ask is "Who is teaching science?" The answer is rather startling. A major in science in college usually means something like 40 semester hours. In a study in Iowa some years ago,[18] researchers discovered that the percentage of high school science teachers with *less than* 20 semester hours (half a major) in their "specialty" was 13 percent for biology teachers, 30 percent for chemistry teachers, and a whopping 63 percent for physics teachers. This is of course not unique to Iowa. Why are these teachers so weak in their own specialties? Did they not intend to teach these courses? Were they pushed into them by school administrators to fill vacant spots? Are the teachers themselves science anxious? Is that why they stopped taking courses in their specialties after only 20 hours?

Whatever the reason or reasons, a large percentage of science teachers can hardly be called specialists in the subject they are teaching. If students discover this, then they are turned off to the course. If students do not discover this, then they pass through the courses blissfully unaware that they may not be getting the most informed or balanced view of the subject matter of science. If the teacher is in fact science anxious, then he or she cannot help but communicate some of this fear to the students. If their only science role model fears science, then there must be something to be afraid of.

What of the well trained science teacher who does not fear science?

Can this person also be a source of science anxiety for the student? Unfortunately, yes. If science is perceived as a very difficult field, then those who go into science are perceived as extremely bright. How does the high school or college science teacher deal with this image? There are two ways. One is to present one's self as a role model to one's students: "I learned science, and I was like you when I was younger, so you can learn science too."

The other way, the way that produces science anxiety, is to present one's self as part of an intellectual elite, a special class of people, and to emphasize by word and gesture that only a few can aspire to this class. Such a teacher gives the message, "I am not like you, and I never was. As hard as you try, you will never be as smart as I am. This material is simply too hard for you."

This kind of intellectual elitism manifests itself in several forms, especially in college. The professor who determines that his or her class will be the filter to "separate the wheat from the chaff," rather than attempting to educate all the students, is producing science anxiety. The teacher who announces on the first day of class that "30 percent of you will have failed or dropped the course by the end of the semester" is the Typhoid Mary of science anxiety. The college science department that makes it clear that it is only interested in educating its majors, that is, those who will become professional scientists, is selling out the rest of the student body and contributing to the bad name that science has acquired. And the large university that runs a pro forma undergraduate major but is really only interested in graduate students is probably impeding science more than its Ph.D. graduates are advancing it.

Even science teachers who love their subject and love teaching may do damage by the way that they react to the attitudes of nonscientists. If an English professor mistakenly expresses contempt for the pursuit of science, it is still totally inappropriate for the science teacher to denigrate the humanities. The science teacher can "sell" science as one of the liberal arts, whereas presenting it as somehow superior to other courses of study can only backfire.

The rate at which science courses are taught is also a source of science anxiety. Here we come up once more against the dual role of the university. There is a great deal of material in a science course. Much of it is necessary for students who wish to take the Medical College Admissions Test or the Graduate Record Examinations. What is the instructor to do? To go too slowly means that a good deal of the material must be left out. To go too fast means that there is no time for questions from the students and no time to absorb the concepts. In general, it appears to me that there is too much material packed into a semester's time, both in the sciences and in the humanities. Perhaps, a serious look at curriculum revision is in order. Short of this step, it is certainly fruitless for the scientist to go too fast. This only heightens the anxiety of the students, since it asks them to build complicated concepts based on ideas that came at them so fast they were not understood. A few concepts learned well make a much better foundation for the study and appreciation of science than many concepts not understood at all. This means some of the ideas that may appear on standard tests such as the MCAT and the GRE may be skipped over. There is no help for this -- except that if the student has come to learn and appreciate science, he or she may be able to read the additional material and comprehend it after the course is over. Certainly the alternative -- a rapid, terrifying course in science -- is no alternative at all. (One might wish to question as we did earlier whether the standardized tests are in fact testing science knowledge in any useful way.)

There is one more way in which science teachers can cause science anxiety. That is by making their courses too easy. Low enrollment figures and hard times have led many universities to introduce popular science courses for students who are neither premed nor science majors. If these courses are strong in intellectual content, if they really teach students to do science, then they are performing a useful function. If however, they are not challenging, if they pander, for example, to concrete reasoning, if they avoid analytical thought, then they are doing more damage than good. Far from giving their students confidence that they can do science, they reinforce the myth that they *cannot* do science; they can only learn *about* science. The student who passes such a course has in actuality achieved little, and knows it -- no new skills, no improvement in reasoning powers, no sense of mastery or accomplishment, and certainly no confidence that he or she can go out and make science-based political decisions. In effect, whatever science anxiety and stereotypes the student brought to the course have been reinforced to the benefit of nobody.

Advances in Popular Science

Everything is not all bleak, however. In recent years there has been considerable progress in science teaching, especially in communicating the wonder of science to people who have no intention of becoming professional scientists. As we discussed in an earlier chapter, science in the nineteenth century was considered a popular subject for "cultured" people to address. Many eminent scientists went out on the popular lecture circuit, not only to supplement their meager teaching salaries, but also to bring science to the people. One should not get the notion, by the way, that only educated middle and upper class Americans were attracted to these lectures. European immigrants in the late nineteenth and twentieth centuries would attend a scientific lecture as their evening entertainment after a grueling day in the sweatshops of New York. Books on science written in lay language, in the tongues of these immigrants, were very popular.

It is only in the last 40 years that science has taken on the aura of forbidden territory. But in the last few years, there have been serious attempts to bring science back to the people. The science lecture for lay people is once again coming into vogue. Scientists are asked to speak on science and related topics (such as energy) at churches, community centers, and other gatherings. Journalists with a science background are in demand,

and more and more articles about science appear in weekly magazines and daily newspapers. Scientific societies maintain public relations offices to determine which scientific advances will be of interest to the press and the public. For example, physicists who are doing pioneering work in some field are asked by the American Physical Society not only to prepare a paper for presentation at a professional meeting but also to prepare a popular version of the paper for dissemination to the media. In addition, they may be asked to hold a press conference to explain in detail the significance of their work.

One important new vehicle for the popularization of science is television. The late Jacob Bronowski presented his lectures on "The Ascent of Man" to a TV audience of millions. Courses based on his book (which is itself based on his TV lectures) grew up in various communities, with presentation of videotapes of the TV series followed by comments and lectures by scientists from the community.

Series on public TV, such as "Nova," have attracted many people to science and have familiarized them with the tasks and the goals of scientists. News specials dealing with politically important topics such as nuclear reactors, genetic engineering, and chemical carcinogens have helped to bring scientific ideas to the public. Science programs for children, such as "3-2-1 Contact," are the most recent additions to this phenomenon.

In addition, there has been a recent boom of popular scientific magazines. For many years *Popular Science* and *Scientific American* dominated a rather limited field. Lately, we see a burgeoning of magazines such as the venerable *Science News*, a short informative weekly report of scientific advances, and *OMNI*, a mix of science fact and science fiction. More recently, the American Association for the Advancement of Science, whose main publication is the professionally oriented *Science*, has entered the popular science race with the bimonthly *Science 80* (and subsequent years). In addition, various public interest and political groups publish scientific information about their particular causes. The Union of Concerned Scientists puts out a newsletter about nuclear power and some of its dangers; the Federation of American Scientists in its "Public Interest Report" concerns itself with the arms race and runaway technology, as does the *Bulletin of Atomic Scientists*.

Charismatic individual scientists have, by their writings and public appearances, helped reawaken the natural curiosity people have about science. Carl Sagan of Cornell University has written a great deal about the planets and about physical science in general.[19] His TV series "Cosmos" has been extremely popular. His appearances on the Johnny Carson Show have given millions of people a close-up look at an articulate, personable scientist who is enthusiastic about his field of study and eager to communicate scientific ideas to lay audiences. Carson himself is an excellent representative of this audience. He not only asks the questions we all wish to ask, but he also expresses our awe and our fear about science.

· Isaac Asimov is another first-rate explainer of science. The usual term is "popularizer" of science, but that, I think, has implications of talking down to the audience. Asimov never does this. He simply explains scientific concepts in terms that the intelligent nonscientist can comprehend. Starting out as a writer of science fiction, Asimov later turned to science writing and has reached millions with his articles and books.[20] Interestingly, both Asimov and Sagan have written on subjects other than science, thus giving the lie to the stereotype of the scientist as a narrow specialist. They see science as one of a number of fascinating intellectual pursuits, the one that by temperament and training has played the largest role in their own lives, but certainly not the only interesting area of study and certainly not the exclusive province of the "gifted few." They and other less famous scientists have made the communication of scientific information to lay audiences their primary mission. The public has responded positively to this reaching out from the scientific community.

In the last decade, as we mentioned earlier, there has been a boom in popular science courses and textbooks for college students who are not majoring in science. This fairly recent trend is in part a reaction to the lower enrollments in some science courses but also in part a recognition that different sectors of the student population need different things from a science course. The science major needs to become adept at the methodology, the tools of the trade, and can therefore spread the content over several semesters. The arts major, on the other hand, needs less methodology and more overview of content. This is not to say that there should be no science skills taught in say, Physics for Liberal Arts Students, lest such a course fall into the category of deceptive advertising mentioned earlier. It does mean that the arts student needs to get a taste of relativity and quantum theory in a one-semester course, where the physics major may not encounter these ideas until the third semester of the physics sequence. Science courses and science texts have become increasingly tailored to the needs of different types of students. This is all to the good. Such courses as Chemistry for Nursing Students can, when properly taught, give the students what they need for their career goals while drawing out their interest in the subject matter because they know that the course is consciously responsive to their needs.

Science Fiction: Science or Fiction?

What role, if any, does science fiction play in making people familiar with, and comfortable with, science? This is not a simple question. Certainly, if we are to believe the recent trend in movies (beginning with *Star Wars*), the space adventure, complete with technological wonders, has great appeal. But does this really have anything to do with science or with people's perceptions of science? Films such as these and TV shows such as "Star Trek" attract large followings, but translating this interest into positive attitudes toward science is another thing altogether.

Western movies and novels are also popular, but we could not say that people learn very much about the history of the American West from them. In fact, the opposite is usually true. These films and stories reinforce a mythology about the West that runs counter to historical fact. Witness fictional treatment of Native American Indians versus the actual treatment of them by settlers and by the government in the seventeenth through twentieth centuries.

A similar case holds for science fiction. Popular films and TV shows tend to be "space operas," the futuristic analogies to the horse operas. The technical gadgetry that they sport should really be considered (if they existed) marvels of engineering. They are clever applications of scientific principles, but they are not themselves science. Furthermore, these space adventures are probably a passing fad and should be treated no more seriously than other forms of escapism.

The actual literature of science fiction deserves a few more words. This literature has been around for more than 50 years in its modern form, beginning with Hugo Gernsback. It has its roots in earlier times, not only in the novels of Jules Verne and H. G. Wells, but even in the speculations of much earlier authors. Anyone who has read or seen the play *Cyrano de Bergerac* probably remembers the remarkable scene in which Cyrano describes several outlandish ways to get to the moon. Edgar Allan Poe was also no stranger to the genre. But does the longevity and popularity of this kind of fiction have any effect on peoples' attitudes toward science? One connection is clear: Both science fact and science fiction activate our sense of wonder about nature. A further connection is the demand of both science fiction writers and popular science writers that we suspend our questions about the details, and simply focus on the results. Science fiction presents us with future societies and futuristic phenomena; we must hold our questions about how they got to be that way. Similarly, popular reports about science give us the remarkable results of research; to give us all the details would be impossible. The problem here is that we see only the surface and never the interior of the problem. Thus while we can marvel at the accomplishments of science (real or fictional), we gain no insight into the process and are thus left as mystified as before. This is actually not a fault of either science fiction or popular science fact reporting. It is not their mission to teach people science. The goal of science reporting is to tell people about results; the goal of science fiction is to entertain and to speculate on the nature of society given certain changes in conditions. In fact, there is a movement in science fiction circles to call the field "speculative fiction," a much more accurate description of what is really going on. Most modern stories of this genre have relatively little science. In fact, their readers are likely to be already proscience. What science fiction may do is to turn young readers on to science. A poll of professional scientists would probably show that a good number of them were science fiction readers in their youth.

The Space Program and Attitudes Toward Science

What effect, if any, did the space program have in turning people's interest to science? The launching of the Soviet satellite Sputnik in 1957 certainly set this nation on a crash program to produce a horde of professional scientists. This program succeeded a bit too well. Many young people who were encouraged to go into science in high school came out of graduate school in the late 1960's and 1970's to find the job market in science was very limited. The space program, in particular, after some golden years culminating in the moon landing in 1969, came to a grinding halt as the nation's economy and its priorities changed.

Only recently has the space shuttle rekindled the public's interest. The average nonscientist is probably positively affected by the space program, although its accomplishments are primarily engineering feats. The science behind spaceflight is simply the application of Newtonian mechanics, which had been well known for 300 years. (This is to be distinguished from the scientific experiments carried out on the shuttle.) Nevertheless, our space flights are seen as an adventure into the unknown and one relatively free of risk to ourselves, at least as compared with scientific advances such as nuclear power, industrial and pharmaceutical chemistry, and genetic engineering. (Whether the money spent on the space program could be better spent on solving earthbound problems such as poverty is a good question -- the fact is, however, that it is not and probably would not be spent, for political reasons rather than for scientific ones.) On balance, the space program probably gives boost to popular interest in science.

Advances in Classroom Teaching of Science

Some substantial progress has been made in the teaching of science, especially at the college level. A fair number of journals are devoted entirely to methods of teaching science. Others include articles about teaching techniques among their topics of interest. Journals such as *Curriculum Review* have sections devoted to the development of science curricula.

In particular, scientists have been devoting much attention to the applications of Piaget's theories. The primary question addressed is whether students can be helped to move from concrete to formal reasoning by teaching science a certain way. Interestingly, there is one very traditional way this has been done -- a way that predates Piaget: the science laboratory course. Scientists have always known that in order for students to really grasp the fundamentals of science, they need hands-on experience. Studies based on Piaget's theories seem to indicate that the laboratory provides the concrete experience that is necessary as a prerequisite for formal operational thought.

A well taught laboratory course can provide a place where the student can "play" with materials and find out how things work. Then, ideally, the underlying principles behind the experiments can be taught formally in a lecture course. If we think of the function of the laboratory in this way, then we can assess the virtues and faults of traditional laboratory courses and make some suggestions about how such courses should actually be taught. One problem has always been that the lab is usually run during the same semester as the lecture. Thus the concrete and formal experiences come simultaneously, which may hinder the learning process.

Another problem is that many laboratory courses have a difficult time finding the right level at which to approach the student -- a problem whose roots lie in the fact that the students are at different levels. Some lab courses are "cookbook": The student follows a rigidly outlined set of procedures without much need or opportunity for independent thought and receives a course grade based on how rigidly he or she adheres to the procedures. At the other extreme is the "open-ended" lab, where the student is confronted with a room full of equipment and told to design and carry out a number of experiments. The first type of lab only focuses on the most concrete thinkers but does not even give them the opportunity to learn by "playing" with the equipment. The second type is appropriate only to those students who are already sophisticated formal thinkers and can probably understand the lecture material without the lab.

Something in between is obviously called for. This is what science teachers have been seeking in the last few years (and even earlier). Many different schemes have been tried, including self-paced labs, programmed manuals, and audiovisual aids. No one "perfect" solution short of individual tutoring has been found, nor should we expect one, since students are at different places in their intellectual development. But it is certainly the case that science teachers have recognized the problem and have been quite creative in attempting various schemes to make the laboratory experience meaningful. (An interesting study indicates that a concrete physics lab course may have helped weak *math* students improve their formal reasoning ability.)[21]

Just as science teachers have attempted to design lab courses that help move the student from concrete to formal thought, so also have they sought lecture style courses or new formats to accomplish the same end. There is some evidence that students who, in the first two years of college, take a concrete science course, can become formal reasoners by the end of it. Not only that, they seem to like it better than the traditional formal course.[22]

There appear to be certain types of courses that can be taught in the high schools or even the junior high schools that help students in developing their scientific reasoning abilities while at the same time becoming familiar with and comfortable with science.[23] Such courses as Introductory Physical Science, Time, Space and Matter, and Earth Science Curriculum Project, all of which emphasized student involvement through investigation of phenomena and deemphasized textbook reading, were able to move junior high school students in significant numbers from concrete to formal thought. Even more significantly, each of these courses was effective for different age levels, showing that a course for 7th graders needs to be tailored differently from one for 8th graders.

A similar approach used in a course entitled Forum for Scientific Inquiry had a positive effect on college students.[24] This course, emphasizing laboratory exploration of phenomena and the availability of

51

science faculty for consultation, and deemphasizing the lecture format and textbook reading, aided college students in moving from concrete to formal thought. These isolated experiments in science learning have yielded positive enough results that science teachers can be optimistic about improving science learning by designing similar courses, courses that deal with students at their own level of development.

We should not construe the success of these courses to mean that the lecture course should be abolished. Once students are on the path to formal reasoning, either by themselves or by taking a course such as those just discussed, there is no reason to avoid the lecture format. The lecture is a way to reach a reasonably large number of students in a reasonable amount of time. We need to consider very specifically what should happen in a science lecture, and this we shall do in Chapter 5. Let us just remark here that there are obviously good and bad science lecturers, and students have a very clear sense of which is which. In Chapter 5 we shall analyze what makes a good or bad science teacher.

Math Anxiety and Science Anxiety

Having discussed the various problems associated with science teaching and learning and indicated some of the recent approaches to alleviating the problem, we now turn to the approach that is the central theme of this book: a direct attack on the anxiety that is behind much of the difficulty that people have learning science.

The story begins some years ago with a study not of science, but of math. Lucy Sells, a sociologist, investigated the number of years of high school math that entering freshmen at the University of California, Berkeley, had taken. She found that 57 percent of entering males had taken four years of high school math, while only eight percent of females had done the same.

Sells then looked into the consequences of differing math backgrounds. She found that without four years of high school math students had effectively barred themselves from the study of the hard sciences, of intermediate statistics (a prerequisite for a career in much of the social science area), and of economics. They excluded themselves from half the possible major fields of study at Berkeley, restricting themselves to humanities, fine arts, and the traditionally female "helping professions," such as guidance, counselling, and elementary school teaching. Their options had been foreclosed before they even got to college.

Sells' report was published in full in 1978.[25] But as early as 1974, when word about the study began to spread, educators at various schools, especially at women's colleges, realized that there was a serious problem that no one had heretofore considered. Sheila Tobias' book *Overcoming Math Anxiety* [26] details the work that has been done since then. One approach was to explore the causes of math anxiety and especially its predominance among women. Earlier studies that purported to show that men were in some ways "naturally better" in math, in the same way that they tend to be, say, taller than women, were called into question. A typical example was the claim that men score better on tests of spatial perception and therefore make better mathematicians. Indeed men do score better on the average on these tests than women do, but it is not clear that this ability is a determinant of general mathematics aptitude. After all, only a limited group of mathematicians -- geometers and topologists -- deals with spatial arrangements. Interestingly, women in physics were at one time urged to go into crystallography, because it supposedly required fine motor skill, like sewing. Yet, crystallography also requires a high aptitude for spatial perception, a characteristic that women supposedly lack compared to men.

At any rate, female and some male educators began to question very closely the evidence purporting to prove women's natural inferiority in math. They were easily able to show that much of the evidence was inconclusive and indeed biased and that in fact women had been socialized to fear and avoid mathematics. Since this had the effect not only of channeling women away from math per se but also of filtering them completely out of the scientific and technical job markets before they even had a choice in the matter, an immediate remedy was needed. Certainly long-range education and political action was necessary to change women's self-perceptions and society's view of them as "nonmathematical." But in the short-range, women already in college needed some assistance to overcome their math anxiety. Thus math anxiety clinics began to spring up on a number of campuses. This not only occurred at four-year women's colleges such as Wellesley and Mills, but also in coeducational institutions noted for emphasis on quantitative studies, such as Harvard and Stanford. Furthermore, although the impetus for these clinics came from women, the clinics were themselves coed, helping men as well as women overcome their fear of and avoidance of math. Looking once again at Lucy Sells' findings, we are struck by the fact that almost half the *men* who enter college (43 percent) are insufficiently prepared for math and science due to their lack of sufficient high school math.

The math anxiety clinics use a variety of techniques, ranging from teaching math skills to engaging in group therapy sessions. Most clinics provide a mixture of various techniques. However, they are not simply math tutoring sessions, since the basic problem that students have is anxiety, not simply lack of skills. In fact, the lack of skills is the result of avoidance, which itself is caused by anxiety. Therefore, math anxiety clinics work on people's self perceptions in order to break down their self-images as "nonmath types." In addition, many clinics also teach relaxation techniques. By practicing relaxing various muscle groups on command, students learn to relax in anxiety producing situations, such as math classes. Some clinics also do systematic desensitization, a technique for conditioning students to respond to math-related situations without anxiety. (We shall have more to say about all this later.) At the same time, of course, the math clinics may teach some skills, since the ultimate goal is to get students back into mathematics so that they will reopen their career options. Mind over Math, a private math-anxiety clinic in New York City run by Stanley

Kogelman and Joseph Warren, teaches no math skills but works entirely on anxiety reduction, and does so quite successfully as measured by the subsequent positive math experiences of its "graduates."

While there is a wide range of activities that occur in math anxiety clinics, one thing they have in common is their recognition that anxiety is the root of the problem and that it must be addressed directly. Their goal is not to produce a generation of professional mathematicians but rather to help people maximize their mathematical potential by removing the barrier of negative self-image and its accompanying anxiety.

What then is science anxiety? Is it related to math anxiety? Is it a single phenomenon or does it appear differently for different sciences? Is it accessible to the anxiety clinic approach?

Math Anxiety and Science Anxiety: Similarities

In the study by Lucy Sells mentioned earlier,[27] it was clear that a poor math background was a barrier to advancement in any technical field, including, of course, the sciences. So there is a connection between doing math and doing science. This may lead us to suspect that math anxiety and science anxiety are closely related to each other. Are they? From our experience with science anxiety, we think the answer is -- yes and no. Yes, there are striking similarities between the two anxieties, but no, they are not one and the same.

First let us distinguish between the various sciences in terms of their connection to mathematics. Biology has the weakest connection (at least biology as taught in the first two years of college). Chemistry and physics are much more closely related to math, using it daily as a tool for understanding the sciences themselves. Let us say at the outset therefore that people who are anxious about learning biology are not so because of math anxiety (although I suspect there are biology students who might have become chemistry students if not for their fear of math). Having said this, let us explore first the similarities between mathematics and the sciences, and then the differences, so that we can distinguish between math anxiety and science anxiety.

The first similarity that strikes a student who attempts a math course and a science course is that both emphasize solving "word problems." We all remember word problems from secondary school: "If Suzy buys two pounds of McKintosh apples at 79 cents a pound and three pounds of Delicious apples at 89 cents a pound and makes them into apple sauce, what is the cost per pound of the apple sauce?" These word problems not only occur more and more often in college math courses, but they are the primary test of a student's knowledge in chemistry and physics courses. Science word problems are the applications of the principles taught in the science course. In fact, they play a larger role in science than in math. A math exam in, say, an algebra course may be a mixture of word problems and direct questions about techniques, such as "Solve the following equations:..."; a science exam is more likely to consist entirely of word problems. Therefore, a student who has become anxious about word problems in high school math courses is likely to carry this anxiety over into both college math and college science. This student's anxiety is both math and science related. For him or her, math anxiety and science anxiety are indistinguishable.

Another similarity between math and science is the use of a mixture of convergent and divergent thinking to comprehend the material. We discussed these modes of thinking in some detail in the last chapter, but we should emphasize again the secondary schools and so-called aptitude tests reward convergent thinking, while mathematics and science require both convergent and divergent thinking. Therefore, one of the skills that entering college students tend to lack is the ability to think divergently. Some students seem to acquire this skill while taking college math and science. Those who do not may wind up math anxious or science anxious or both.

Science and math skills look very similar when we compare them to the skills necessary for art and literature. Textbooks in math and science are, for example, to be read much more slowly than are humanities texts. Problem solving is peculiar to math and science. Mathematicians and scientists operate under the assumption that nature has an order that it is their task to uncover; humanities scholars, writers, and artists do not make this assumption at all. This difference in approach and in necessary skills leads not only to the myth that there is an "artistic" and a "scientific" mind, but also to the myth that those who are talented in math are automatically talented in science, and inversely, those not talented in one are not able to do the other. This myth is as false as the former one. We might again use as our example Einstein, who, when he first began to conceive of his theory of general relativity, considered by many to be the greatest intellectual achievement in physics, realized he did not comprehend sufficient math to work the theory out. He had to go to his friend the mathematician Marcel Grossman and learn the necessary math from him -- a task that took several years and that Einstein did not describe as painless by any means.[28]

Einstein's physical intuition was superior to his mathematical ability: The two are not one and the same. But many people think they are, especially when compared with the arts and literature. This belief tends to blur the distinction between math and science, and fear of one may quickly produce fear of the other.

Another similarity between math and science is the use of formal reasoning. Both math and science require that students have some experience with concrete reasoning before they can move to the stage of formal reasoning. In science, this can be done by giving students hands-on laboratory experience in high school, for example. In mathematics, concrete reasoning is developed by giving students practical experience with numbers (arithmetic) and with spatial orientation (geometry). However, whether or not these attempts are actually able to prepare students to move on to formal reasoning is the problem. We discussed this in the last chapter with regard to science; the situation is a little different for math. Since college science and college math courses are taught on the basis of formal reasoning, those students who are not ready to reason in this way will probably encounter difficulty in both the sciences and in mathematics. These students (who are likely to be more anxious about chemistry and physics than about biology) may exhibit both math and science anxiety.

Both math and science also act as job filters, in particular against women. The situation in math as outlined by Lucy Sells has its counterpart in the sciences. Women make up a minority of all freshman science majors, with the numbers decreasing in subsequent college years. Biology has the least disparity, with chemistry next and physics the worst. This is not to suggest that physicists actively discriminate more than chemists or biologists, but rather that people perceive physics as the hardest and the most mathematical of the sciences. Young girls, still trapped by societal myths and lacking female scientist role models, respond accordingly by avoiding physics the most. Thus both math and science anxiety act as sex-selective career filters.

It seems apparent that a similar selection process operates against disadvantaged minorities.[29] This is a topic worthy of a detailed study by social scientists and immediate remediation. In the meantime, math and science anxiety clinics must be especially sensitive to discrimination against women and minorities and look for race and gender-relative negative self-messages from these students in the Clinic. One striking example of this was the case of a black woman in our Clinic who had been unable to do either math or science since grammar school, when her white teacher had told the class of inner city youngsters that they could not expect to do well on standardized tests since they were in competition with suburban whites.

Math Anxiety and Science Anxiety: Differences

Having outlined the similarities between math and science anxiety, let us now consider the differences. The first "difference" is simply a fact that we observe: Most students who come to our Science Anxiety Clinic usually are doing fine in math. They report little or no anxiety in their math classes, but a great deal of anxiety in studying science. Conversely, we presume that there are math anxious people who can do well in science, although this is more likely in biology than in chemistry or physics, where math is used as an everyday tool. In fact, we suspect, as mentioned earlier, that some students choose the science they will study on the basis of its math content. Nevertheless, it is clearly the case that one can be science anxious without being at all anxious about math. In fact, we administered both math-anxiety and science-anxiety tests in the form of questionnaires to the students who signed up for the Clinic. There was little correlation between the two phenomena. However, those students who did register math as well as science anxiety at the beginning of the Clinic were able to reduce both as a result of the Clinic procedures, no doubt because the procedures are similar for math and science anxiety clinics.

One very significant difference is that science, unlike math, is based on experiments. The theory taught in science lectures derives its validity from results found in the laboratory. Mathematical theories, on the other hand, while "true" in the same sense that scientific theories are true (e.g., testable by independent researchers), are in a sense self-contained: What you see in the math lecture is what you get. What you see in the science lecture is based on something external to theory, namely, the experiment. It is as if an important component of science is absent from the teaching process. This is partly rectified by requiring a laboratory course along with the science lecture -- assuming the two are in synchronization with each other, which is not always the case. So, in a sense, the science lecture suffers from the same drawback as the popular science program on TV: It gives a body of results, while unintentionally concealing the difficult process that went into obtaining them. This is not to say, of course, that doing creative math is easy, but the way math is done is more closely approximated by the math course than the way science is done is approximated by the science course.

This difference has a number of consequences. The most obvious is that science anxiety has a component that is absent from math anxiety; namely, laboratory anxiety. Students who may be able to function well in lecture courses may be highly anxious when faced with a morass of equipment and asked to perform an experiment to test the theories they learned in the lecture. The Science Anxiety Clinic has to address this issue directly for those students who have joined the Clinic because they are terrified of lab.

Another consequence of the experimental nature of science as contrasted with math is a bit harder to explain. We may state it as follows: Problem solving in science is viewed by students as somehow different from problem solving in math. Part of this stems from the fact that math is teaching students certain techniques, while science is forcing them to apply these techniques. Let us take the following example. A college student is taking a course in calculus and a course in physics. The topic covered in calculus is "differentiation." (You don't need to know calculus to understand this example.) Differentiation is a mathematical operation, like addition or taking square roots. The job of the math professor is to teach the technique so that the student may apply it not only in pure math but in applications of math, such as physics. Therefore, once the student has learned, or supposedly learned, how to perform the operation of differentiation, the math test will ask two types of questions:

1. Differentiate the function x = 4 at^2.

2. What is the velocity of an object whose position varies as the square of the time?

The first of these questions is a straightforward request to the student to demonstrate mastery of the technique. The second question is a word problem involving the student's knowing that velocity is the result of differentiation of position, but it is still an application of the same technique.

The student who is not math anxious will recognize this by the context and will probably be able to work the second as well as the first problem.

Now consider the same student in a physics course. Here the student is studying about motion, including such ideas as position, velocity, acceleration, mass, and force. This student when given a problem similar or identical to question two, but in the context of a physics exam, may be too anxious to do it. How is that possible? The physics course is drawing not on a limited area of abstract knowledge, such as differentiation, but is instead using differentiation rather casually to explain the results of physical experience. So the student first of all does not get as good a clue from context about what is being asked for, since skills other than differentiation are needed to describe motion. Second, and very important, the student perceives the math course as covering separate, well compartmentalized topics while he or she *perceives* the physics course as describing all the physical universe using mathematical models. The science-anxious physics student assumes that there is something about nature that others have grasped but that he or she has not. The student goes looking for the forest while ignoring the trees and consequently fails to solve the problem.

This familiar feeling that "something is missing" characterizes the science-anxious student who may not be math-anxious. In *Overcoming Math Anxiety*,[30] Sheila Tobias describes a related phenomenon in math anxiety; namely, the focusing on irrelevant details as a way to avoid doing the math. To the purely science-anxious student, math usually seems free of such irrelevant detail, but he or she is always suspicious that science contains important details that were missed while trying to understand the physical model of nature that has been presented.

It is essential to realize that this perception about science is false. Yes, there is a big "forest" out there -- the universe -- but the way you learn about it is by going from tree to tree. The science-anxious student tends to read more into the content of science than there may be; if a question on an exam looks easy, then the student assumes that he or she has missed the point. This is not to suggest that science is easy or one-dimensional. Certainly, as one progresses through the study of science, one sees more and more facets of nature, and indeed one begins to see the "forest." However, this is not something that happens in the first year of college science. The science-anxious student seems to assume that everyone else is grasping the "big picture." This erroneous belief lies at the heart of much of the fear of science.

We can complete our survey of the differences between math and science anxieties by asking: How do people view the relevance of math and science to their lives? It is clear they view math as more immediate and relevant. Math anxiety includes such issues as inability to balance a checkbook or add up the cost of groceries or figure out your income tax. Anyone who has gone to school has had contact with math. It is familiar, whether you like it or not. Not so for science. It is possible, thanks to past generations of science-anxious educators, for students to get high school diplomas with minimal or no contact with science. But everyone is affected by science and its consequences: medical techniques and drugs; chemicals, both beneficial and harmful to us and to the environment; the internal combustion engine; products of physics, such as nuclear energy and nuclear bombs, lasers, transistors, and integrated circuitry; all affect our daily lives. Yet most people have managed to avoid learning anything about these technological wonders and dangers because they are afraid of the science on which they are based.

Science, then, is omnipresent, and in a more threatening way than math is: An unbalanced checkbook simply does not produce the same dangers as an unbalanced nuclear reactor. Paradoxically, and most unfortunately, this fear of the consequences of science seems to produce a fear of learning science, which places people even more at the mercy of science and its technological offshoots. The student who is beginning to study math and science, then, brings the psychological burden to the subject and shies away from science at least as much and probably more than away from math. These kinds of generalized fears shared by many must be addressed by those who hope to lessen science anxiety and thus to produce a scientifically literate populace, just as the related but different fears of math have been addressed by people who work with the math anxious. The Science Anxiety Clinic does precisely that.

The First Science Anxiety Clinic

In 1974, I was teaching physics and astronomy at a state university. One of my teaching duties was a course in introductory physics for nonmajors. The students in the course were majoring in biology, in the social sciences, and, in some cases, in the liberal arts. The students seemed bright and sufficiently prepared by their high schools to grasp the material of the course.

One of the duties of a professor is to provide office hours, when students can come in with questions that they may have about the course. A number of students used to drop by during my office hours complaining that they simply were unable to catch on, to master the skills necessary for doing well in physics. Now the basic skill in physics is to comprehend a concept and then apply it to the solution of problems. For example, a student would need to understand the relationship between the force applied to an object and its response to this force, namely, its acceleration. The test of whether the student had indeed grasped the concept would be whether he or she could work out numerical problems involving various objects accelerating under the influence of a force.

The students who came into my office often seemed unable to master this skill. At first I assumed they were simply the weaker students. I spent a good deal of time working through problems slowly, step by step, with them. A few days later they would return with the same or similar problems. When I checked a little more closely, I discovered that they were not in general poor students; in fact, they seemed to do rather well in other courses. But science seemed to be throwing them. Was it then that they simply didn't have a "scientific mind?" I didn't really believe that such a thing existed. After all, we were not asking them to carry out original research in science. All we were asking was that they master some basic skills, just as they were doing in, say, their English composition or psychology classes. Why were they having so much difficulty?

At the same time I was asking myself these questions (and the answer didn't occur to me for awhile), I was beginning to notice other features of the students' behavior, to pick up signals from their body language. They seemed clearly agitated. I began to question them about this. They freely admitted a great deal of anxiety. In fact, some of them could trace their anxiety to things other than science, for example, problems at home. Yet the effect was poor performance in science. No matter how bad things were 'at home, they were still able to pass that English composition class. But the science course was the first casualty of their anxiety, and in fact was the focus of this anxiety. What to do? Clearly, giving them extra problems to solve was not going to help. The last thing they needed was more work. Individual counseling was a possibility, but this approach had various shortcomings. One was the unreasonable hesitation that students feel about going to the school counseling center. They had a problem they perceived as academic. To suggest a psychological solution was not appropriate or useful. Second, if all the students who were anxious about science agreed to go for counseling, the counseling center would be bursting at the seams. Third, and most important, the problem was one that was limited to their science experience and as such should be accessible to a joint attack by a psychologist and a scientist, rather than by either working alone.

At the time these ideas began to jell in my mind (I had moved to Loyola by that time), I learned about the then-very-new math anxiety clinics. I found out about them by chance, through a friend who is an artist. At the conclusion of a heated argument about whether everyone could find something exciting about math and science (She: no! I: yes!), she admitted her own fear of mathematics and science and related some horror stories about her experiences trying to learn these subjects from unsympathetic teachers. A few days later, she showed me an article about math anxiety clinics. I realized that this idea could be adapted to science anxiety. Of course, there were some problems. There are many sciences: Could we deal with the problem in general terms or would we need a biology clinic, a chemistry clinic, and a physics clinic? Was science anxiety a separate phenomenon from math anxiety or was science anxiety just a consequence of an earlier math anxiety? Was science anxiety simply test anxiety? That is, were the students actually learning the material but simply freezing up on exams? These and many other questions occurred to me, But -- "nothing ventured nothing gained." I approached the Loyola Counseling Center with my idea for a Science Anxiety Clinic.

I had a general model in mind. The clinic should be limited in time, say seven or eight weeks, an hour or two a week. It should consist of groups of not more than 10 students. It should be led by a scientist (me) and a psychologist. They should be of opposite gender, since I predicted that many more women than men would join the clinic and I felt they should have a role model to relate to. The clinic should employ all the techniques necessary to reduce anxiety about science, including development of science learning skills, discussion of past bad experiences in sciences, changing of negative self-images, relaxation and desensitization techniques, and anything else we could think of.

The Counseling Center approved my idea. In 1976 I began the first Science Anxiety Clinic with Marian Grace, a Counseling Center staff psychologist, as my co-leader. The Clinic was announced in various introductory physics courses. Students signed up voluntarily. Dr. Grace and I ran the first clinic for seven weeks, beginning a few weeks into the semester and ending just before final exams. Positive feedback from the students involved encouraged us to continue the project. The next year we applied for and received a Mellon Foundation Grant, administered through Loyola, and we expanded the Clinic to three groups. Under Dr. Grace's supervision several graduate students in counseling worked with me in the groups.

In addition, a research project was begun to measure the effect of the Science Anxiety Clinic on actually reducing anxiety. This project was carried out by Rosemarie Alvaro; it became her Ph.D. dissertation. She developed a questionnaire to ascertain how anxious students felt about science before and after their experience in the Clinic. She also made a physiological measurement of their response to science situations, by measuring their muscle response (electromyography) to science-related stimuli. She found that the Clinic was extremely effective in reducing anxiety. Virtually all the students who went through the Clinic showed significant anxiety reduction, compared with a control group of students who had applied to the Clinic but whose schedules made it infeasible for us to accommodate them.

A second study, carried out more recently by Joe Hermes as his Ph.D. dissertation, corroborated Alvaro's findings, as well as demonstrating that an individual Stress Management Program, emphasizing biofeedback, could also be effective in science anxiety reduction. Hermes concluded that the Stress Management Program could be used with students suffering from generalized anxiety, while the Science Anxiety Clinic could best benefit specifically science-anxious students and/or students with weak science study skills.

These positive results encouraged us to continue the project. Today we are a regular part of the Counseling Center activities. Unfortunately, we are able to accommodate only about half the students who apply, due to scheduling difficulties and simple lack of time. The Clinic has received wide press coverage from *Newsweek* to the *London Times,* Swiss National Radio, and even the *Netherlands Journal of Materials Research.* We have received several hundred letters asking for information about setting up science anxiety clinics at colleges and high schools across the country.

Science Anxiety Questionnaire

Date: Name:

The items in the questionnaire refer to things and experiences that may cause fear or apprehension. For each item, place a check mark on the line under the column that describes how much YOU ARE FRIGHTENED BY IT NOWADAYS.

	Not at all	A little	A fair amount	Much	Very Much
1. Learning how to convert Celsius to Fahrenheit degrees as you travel in Canada.					
2. In a Philosophy discussion group, reading a chapter on the Categorical Imperative and being asked to answer questions.					
3. Asking a question in a science class.					
4. Converting kilometers to miles.					
5. Studying for a midterm exam in Chemistry, Physics, or Biology.					
6. Planning a well balanced diet.					
7. Converting American dollars to the English pound as you travel in the British Isles.					
8. Cooling down a hot tub of water to an appropriate temperature for a bath.					

	Not at all	A little	A fair amount	Much	Very Much

9. Planning the electrical circuit or pathway for a simple "lightbulb" experiment.

10. Replacing a bulb on a movie projector.

11. Focusing the lens on your camera.

12. Changing the eyepiece on a microscope.

13. Using a thermometer in order to record the boiling point of a heating solution.

14. You want to vote on an upcoming referendum on student activities fees, and you are reading about it so that you might make an informed choice.

15. Having a fellow student watch you perform an experiment in the lab.

16. Visiting the Museum of Science and Industry and being asked to explain atomic energy to a 12-year-old.

17. Studying for a final exam in English, History, or Philosophy.

18. Mixing the proper amount of baking soda and water to put on a bee sting.

	Not at all	A little	A fair amount	Much	Very Much

19. Igniting a Coleman stove in preparation for cooking outdoors.

20. Tuning your guitar to a piano or some other musical instrument.

21. Filling your bicycle tires with the right amount of air.

22. Memorizing a chart of historical dates.

23. In a Physics discussion group, reading a chapter on Quantum Systems and being asked to answer some questions.

24. Having a fellow student listen to you read in a foreign language.

25. Reading signs on buildings in a foreign country.

26. Memorizing the names of elements in the periodic table.

27. Having your music teacher listen to you as you play an instrument.

28. Reading the Theater page of *Time* magazine and having one of your friends ask your opinion on what you have read.

	Not at all	A little	A fair amount	Much	Very Much

29. Adding minute quantities of acid to a base solution in order to neutralize it.

30. Precisely inflating a balloon to be used as apparatus in a Physics experiment.

31. Lighting a Bunsen burner in the preparation of an experiment.

32. A vote is coming up on the issue of nuclear power plants, and you are reading background material in order to decide how to vote.

33. Using a tuning fork in an acoustical experiment.

34. Mixing boiling water and ice to get water at 70 degrees Fahrenheit.

35. Studying for a midterm in an History course.

36. Having your professor watch you perform an experiment in the lab.

37. Having a teaching assistant watch you perform an experiment in the lab.

38. Focusing a microscope.

	Not at all	A little	A fair amount	Much	Very Much

39. Using a meat thermometer for the first time, and checking the temperature periodically till the meat reaches the desired "doneness."

40. Having a teaching assistant watch you draw in Art class.

41. Reading the Science page of *Time* magazine and having one of your friends ask your opinion on what you have read.

42. Studying for a final exam in Chemistry, Physics, or Biology.

43. Being asked to explain the artistic quality of pop art to a 7th grader on a visit to the Art Museum.

44. Asking a question in an English Literature class.

[1]A. Ahlgren and H.J. Walberg, "Changing Attitudes Towards Science Among Adolescents," *Nature* 245:187-190, 1973.

[2]D.C. Beardslee and D.D. O'Dowd, "The College-Student Image of the Scientist," *Science* 133:997-1001, 1961.

[3]L. Hudson, "The Stereotypical Scientist," *Nature* 213:228-229, 1967.

[4]A. Roe, *The Making of a Scientist*, New York: Dodd, Mead, 1953.

[5]A.J. Cropley, "Divergent Thinking and Science Specialists," *Nature* 215:671-672, 1967; Cropley and T.W. Field, "Achievement in Science and Intellectural Style," *Journal of Applied Psychology* 53:132-135, 1969; J.F. Feldhusen, et al., "Anxiety, Divergent Thinking, and Achievement," *Journal of Educational Psychology* 56:40-45, 1965.

[6]R.G. Fuller, R. Karplus, and A.E. Lawson, "Can Physics Develop Reasoning?" *Physics Today* 30:23-28, February 1977.

[7]J.W. Renner et al., *Research, Teaching, and Learning with the Piaget Model*, Norman, OK: University of Oklahoma Press, p. 66, 1976.

[8]J. Piaget, *The Psychology of Intelligence*, London: Routledge and Paul, p. 148, 1967.

[9]Renner et al., *op. cit.*

[10]G. Barnes and G.B. Barnes, "Students' Scores on Piaget-Type Questionnaires Before and After Taking One Semester of College Physics," *American Journal of Physics* 46:807-809, 1978.

[11]G. Barnes, "Scores on a Piaget-Type Questionnaire vs. Semester Grades for Lower Division College Physics Students, " *American Journal of Physics* 45:841-847, 1977.

[12]H.D. Cohen et al., "Cognitive Level and College Physics Achievement," *American Journal of Physics*, 46:1026-1029, 1978.

[13]D.F. Griffiths, "Physics Teaching -- Does It Hinder Intellectual Development?" *American Journal of Physics* 44:81-85, 1976.

[14]A.B. Arons, "Cultivating the Capacity for Formal Reasoning: Objectives and Procedures in an Introductory Physical Science Course," *American Journal of Physics* 44:834-838, 1976.

[15]Renner et al., *op. cit.*, pp. 76-77.

[16]Griffiths, *op. cit.*

[17]Renner et al., *op., cit.*, pp. 90-109.

[18]P.W. Tweeten and R.E. Yager, "Content Preparation of Science Teachers in Iowa," *School Science and Mathematics* 68:824-833, 1968.

[19]See for example, C. Sagan, *Broca's Brain: Reflections on the Romance of Science,* New York: Random House, 1979; Sagan et al., *Planets,* New York: Time, Inc., 1966.

[20]See for example, I. Asimov, *Asimov's Guide to Science,* New York: Basic Books, 1972.

[21]R. Gerson and R.A. Primrose, "Results of a Remedial Laboratory Program Based on a Piaget Model for Engineering and Science Freshmen," *American Journal of Physics* 45:649-651, 1977.

[22]J.W. Renner and W.C. Paske, "Comparing Two Forms of Instruction in College Physics," *American Journal of Physics* 45:851-860, 1977.

[23]Renner et al., *Research, Teaching, and Learning.*

[24]Ibid.

[25]L. Sells, "Mathematics as a Critical Filter," *The Science Teacher* 45:28-29, February 1978.

[26]S. Tobias, *Overcoming Math Anxiety,* New York: W.W. Norton, 1978.

[27]L. Sells, op. cit.

[28]B. Hoffman and H. Dukas, *Albert Einstein: Creator and Rebel,* New York: Viking Press, 1972.

[29]L. Sells, *Curriculum Update,* California School Boards Association, No. 41, June 29, 1979; also private communication.

[30]S. Tobias, op. cit., pp. 124-125.

4

Overcoming Science Anxiety

The last chapter briefly introduced the Science Anxiety Clinic. Chapter 6 discusses the Clinic in detail. Unfortunately, at this time, there are relatively few such clinics. Therefore this chapter outlines ways in which the science-anxious person can begin to combat his or her fears without having to find a Science Anxiety Clinic in the area.

What are the goals of this chapter? First, to summarize the various skills necessary for learning science. Second, to list the various forms that anxiety takes and the negative self-statements and irrational beliefs that underlie the anxiety. My hope is that the science-anxious reader will be able to identify some of his or her self-statements from the list that is provided. Then, the next step is to find coping statements. This is admittedly where we are on shaky ground, for, as any good psychologist knows, you cannot tell someone what changes he or she needs to make; they must find this out through a personal process of self discovery. For our purposes, this means that each person should work out his or her own coping statements. This is precisely what we do in the Clinic. In this chapter, that is obviously impossible, so we can do no more than provide a discussion of various coping statements and hope that seeing the ones that apply will plant a seed of self discovery in the science-anxious reader.

Finally, this chapter will consider the "science hierarchy." We shall then discuss briefly the relaxation and visualization processes that we use in the Clinic and tell the reader where various cassettes for relaxation are available.

For this chapter to be of any use at all, the reader must become an active partner, not simply a passive recipient of knowledge. In fact, a good way to read this chapter is the same way you should read a science textbook: with pencil and paper in hand, stopping frequently and thinking of examples of how the material is applicable, in this case, to your own personal form of science anxiety. Both the person who would like to study science but is afraid and the person who is already studying science and is anxious can benefit from this chapter.

Science Skills

Although the skills needed to learn science are somewhat different from the skills needed to learn humanities, there is nothing mysterious about them. However, it is often the case that science teachers do not explicitly state what these skills are, but assume that students will pick them up by example and implication. This assumption is not correct.

Therefore, let us now have a look at what these skills are and let us say a little about how they can be developed. Some of the skills are related to mathematics and are the same ones that are taught in math anxiety clinics. Others are specific to science and indeed specific to particular sciences and not to others.

Reading Science

There are various types of science writing, from popular science articles in newspapers and magazines to more detailed articles in *Science News* and *Scientific American* (articles that, however, require no prior knowledge of a subject) to textbook writing and, finally, to articles in scientific research journals. Each requires a different kind of reading style and different expectations about what can be learned from the content.

We shall not deal with research journal reading here. Of the other three, the easiest articles are those in the popular press. However, they have the least content and can sometimes be unintentionally misleading. In the first place, they are written in journalistic prose style. That is, the author's intention is to communicate the ideas to the reader in one fairly fast reading, just like nonscience articles. But, as we have emphasized earlier, you cannot learn science this way. What then is the popular article imparting to the reader? It is imparting the results of some scientific research. It cannot in general give the reader much insight into the process by which the research was conceived and carried out, nor about the logical train of thought underlying the research. We may read that an unmanned spacecraft is studying Jupiter; we will not learn how the particular problems to study were devised or why they are considered the essential ones or indeed how they fit into the general range of astronomical problems. The same can be said for science reporting on television: one sees results, but one does not get to follow the process of science.

There are two problems with this approach. The first is that it gives a somewhat misleading view of science. It appears as if the scientist can by magic choose the "right" problem to study and then quickly find a solution.

Not so. Both the choice of problem and its solution are difficult tasks with no assurance of success -- that's why it's called research. The second difficulty, especially for the person who wishes to understand the logical processes of science, is that the popular article may give the reader the impression that it contains more than it actually does. In particular, the article may look as if it is *explaining* something, when in fact it is only *describing* something. The reader, not understanding the logical process behind the scientific result, may then assume that he or she has missed something in the article, when in fact that something, the *explanation*, is not really present. Thus is born a negative self-statement: "If I can't even understand this popular article, I must really be completely incapable of comprehending science!" The only effective way to combat this self-statement is to read the popular article the way it should be read: for results but not for explanation. Do not expect more than is there.

The next level of science writing is the more serious article in the science magazine. Here, the author's purpose is indeed to give the reader some insight into the process of science as well as to present results. Although the author makes no demands that the reader have any prior background, except perhaps for a good general education, the author does demand a good deal of concentration on the part of the reader. Nobel Laureate Steven Weinberg, in his popular book about the origins of the universe, *The First Three Minutes*,[1] makes this demand explicitly and articulately:

"However, this does not mean that I have tried to write an easy book. When a lawyer writes for the general public, he assumes that they do not know Law French or the Rule Against Perpetuities, but he does not think the worse of them for it, and he does not condescend to them. I want to return the compliment: I picture the reader as a smart old attorney who does not speak my language, but who expects nonetheless to hear some convincing arguments before he makes up his mind."

Thus as you read this kind of serious science writing for the nonscientist, you must read slowly, chew over each idea, and even go back and forth from earlier to later paragraphs to see if you follow the logical connections. You may not actually need pencil and paper in hand, since the articles may be fairly qualitative. But you will need to read slowly. When you have finished, however, you will, if the writing is good, have acquired more insight into how science is actually done than if you had read an article on the same topic in a newspaper. At all times, by the way, you should be assessing whether you think it is or is not good writing. Science writing like any writing is not of uniformly good quality, and it demystifies science to keep this in mind.

Finally, what about reading a science textbook? This must be done slowly, with pencil and paper in hand, and must be done more than once. Ideally, a chapter should be read once before it is covered in a lecture, once during the time it is being covered, and once after it has been covered. Even before reading the chapter, you should read the chapter headings and subheadings just to get an overview of the material. At each step in which a new concept is presented, you must think it over, see if it makes sense to you, and work out an example if it is possible. Often in science texts the author will illustrate a concept by working an example. You should not only read the example for comprehension, but then work it for yourself without looking at the author's solution. When you can do this, and only then, can you feel sure that you have mastered the concept.

Keep in mind that it is not possible to skim the material for content. Skimming may provide a general overview, but it cannot provide the comprehension of the logical processes on which the material builds.

It is useful to look at more than one text's treatment of the same material. Find out what the instructor's second and third choices would have been for the course text and take a look at them. Often the auxiliary text will provide just a slightly different viewpoint, but one that makes more sense to you than the course text's treatment. Science, like art, is multifaceted. Each time you look at the same bit of work, you may see something different. So it is useful to read how various authors of science texts view the same concept. In fact, it is similar to reading two historians' views of the same series of events -- except that the scientific data are reproducible.

The Science Lecture

The purpose of the science lecture is to complement the text -- to give the students the teacher's view of the material. The teacher does this in several ways: by presenting a logical overview of the material, by emphasizing certain parts of the material over other parts, by presenting related material not explicitly in the text, and by giving examples of the application of the concepts presented. The science lecture, like any serious lecture in any serious course, is therefore not really self-contained. It requires the student bring some knowledge to the lecture; specifically, knowledge gained from the text. The difference between the science and the nonscience course is that in the latter it is often possible to follow the lecture and get the gist of it before reading the text (although this is not a good idea, and in some cases, such as a discussion of a work of literature, is useless). In a science lecture getting the gist is well nigh impossible without having read the text. Furthermore, the lecture format itself may be deceptive, albeit not intentionally. The instructor may make the material look easy, and the student may think that it is possible to grasp the material without the text. This is in general not so. The text provides the basis for the lecture, even if it doesn't appear that way. In fact, the text is the basis for lectures that are themselves digressions from the text material. Without knowing the text, the student cannot get any feeling for why the instructor considers a particular digression important.

The lecture also makes possible feedback from student to instructor in the form of questions. But the student must be prepared beforehand in order to ask these questions. The form this preparation takes is to read the text and decide which sections you do not understand. Then the lecturer's role is to clear up these sections when he or she reaches that point in the lecture. If the idea is still unclear, then you need to ask a question. With this approach, the questions tend to be precise, well formed, and explicit. It becomes easier for the instructor to focus in on the exact nature of the student's problem and to give a clear answer.

Naturally, you cannot always anticipate what the questions will be. The lecture itself may very well raise questions on the spot, questions that do not stem from the text. These questions should also be asked, but even they can be asked more precisely if you have studied the text.

While listening to a lecture, you must be asking yourself, "What is the instructor emphasizing? What is he or she leaving out? Why is this material important? How does this idea follow from the previous one? How is this treatment different from the text's treatment of the same material?" These questions will help you understand the lecture better and also help you make clearer notes.

How, in fact, should you take notes? What should you write down? What should you leave out? What if you get lost in the middle of an idea? Should you stop and think about it or keep writing?

If you keep in mind that the lecture complements the text, then the answers to these questions are not hard. Write down the points of emphasis, write down the logical steps the instructor uses to illustrate an idea or prove a statement, but don't try to get down every word that is uttered. If you come to a place where you lose track of an idea, and you cannot quickly frame a question for the instructor, mark the spot in your notes and keep writing. If you stop for a few minutes to think about the idea, you will lose more of the lecture, and the chances are that in the heat of the moment you will not be able to grasp the point you have missed anyway. Better to know where the point was in your notes and go back to it after class. Then you can look at the text, ask someone in the class, or go to the instructor and ask. Keep in mind that the whole purpose of note-taking is to look at the notes later, put the information together, and make sense of it. Even the parts of the lecture where you were not lost have aspects that you did not grasp while hearing them for the first time, so they are not so different from the part where you lost track.

All this discussion has assumed a competent instructor, one who is willing to help students and to take questions as they arise. Obviously this is not always the case. Poor instruction, as we said earlier, is one of the causes of science anxiety. We will have more to say about this in the next chapter. At this point we can only point out that you cannot change poor instruction by yourself, although there are some things that students can do in the case of gross incompetence. What is important is that you look for ways around the poor instructor, such as working with other students (always a good idea in science, even with good instructors), sitting in on other sections of the course taught by good instructors, and seeking tutoring. What you must not do is turn the instructor's incompetence or lack of willingness to help into a negative self-statement such as, "I must be stupid if he or she won't answer my questions." It is the student's confusion of self worth with performance that lies at the heart of much science anxiety.

A very good way to see if you have grasped a concept is to lecture on it yourself. This is another reason why it is good for students to work together on science. After each has read the exam and attempted the homework separately, they can get together and help one another. As they lecture to each other on this or that concept, not only the listener but the speaker will gain more insight into the ideas being presented. It is no accident that professional scientists tend to work in groups. The breadth and depth of the ideas involved militate in general against lone wolves. By following this model, students of introductory science can help themselves learn more, and faster.

Word Problems

Without looking at the answers below, try to work the following two problems:

Problem 1. The universe started as a "Big Bang" explosion. The galaxies that received the greatest velocities at that time have moved the farthest. Today we measure a galaxy to be 2×10^{21} kilometers away and receding from us at 10^{11} kilometers per year. Assuming that the velocities of the galaxies have remained constant since the Big Bang, calculate the age of the universe.

Problem 2. Detroit is 300 miles from Chicago. How long does the trip from Chicago to Detroit take, at a constant speed of 60 miles per hour?

Let us first see how to work the second problem, the one that looks simpler.

A good way to solve such science (or math) word problems is to make three columns headed What We Know, What We Want, and What Relates Them. In Problem 2, we are given the distance between Chicago and Detroit and the speed at which the car is traveling. We are asked to find the time the trip takes. The thing that relates them is some connection between distance, velocity, and time expressed in equation form. So the format for the solution of Problem 2 is:

Solution 1

What We know	What We Want	What Relates Them
$d = 300$ miles $v = 60$ miles per hour	$t = ?$	$d = vt$

We introduce symbols for the known and unknown quantities and write the relation between distance d, velocity v, and time t as the equation $d = vt$; namely, distance = velocity x time. We then solve for time:

$$t = \frac{d}{v} = \frac{300}{60} = 5 \text{ hours}$$

Many people are able to work this problem in their heads. However, outlining this logical procedure teaches you how to set up more complicated problems. One of the erroneous beliefs that students have about science is that if you only know the formula you can solve the problem. This distracts you from the first two critical steps: deciding what you are given and what you are asked for. Memorizing formulas is no trick; knowing how to set up the problem so that the formulas may be applied is what is crucial.

Once having outlined the general scheme of solution, we may turn to the solution of the other problem.

Solution 2

What We Know	What We Want	What Relates Them
$d = 2 \times 10^{21}$ kilometers $v = 10^{11}$ km per year	$t = $?	$d = vt$

Solution:

$$t = \frac{d}{v} = \frac{2 \times 10^{21}}{10^{11}} = 2 \times 10^{10} \text{ years}$$

This is the age of the universe.

Note how similar the two problems are. Both use the same assumptions -- car and galaxies travel at constant speeds -- and the same math. If you are like most people, you are more comfortable with the car problem. Why this is so we shall see later (p. 142). Right now, the important thing to remember is that working a science problem is not simply searching your "memory bank" for the right formula, but rather, writing down what you are given, what you are asked to find, and only then the question or equations that relate them. Let us give one more example here. This time, we will make it a little more complex. There are several steps involved in getting from what we are given to what we are asked to find.

Problem. What force is needed to accelerate an object of mass 2 kg from rest to 30 meters per second over a distance of 45 meters?

First we list <u>What We Know</u>

Object starts "at rest":   speed $v_1 = 0$ meters/sec
Object reaches speed v_2 - 30 meters/sec
$M = 2$ kg
$d = 45$ m

Now we state <u>What We Want</u>

Force F on object

This type of problem usually appears in the first month or so of a basic physics course. By that time the students have learned that acceleration can be found from changes in velocity over a period of time or over a certain distance. They have also found that to accelerate an object, you need to apply a force. And they have seen a number of equations that relate these concepts. When they try to relate what is known and what is asked for in the problem just given, they discover that this cannot be done in one simple equation, but needs two:

<u>What Relates Them</u>

Acceleration $a = \dfrac{v_2^2 - v_1^2}{2d}$

Force $F = ma$

They have chosen the first equation from a number of equations relating distance, velocity, and time by eliminating those that are not asked for; to whit, any involving time. Thus the first equation relates acceleration to velocity and distance traveled, with no mention of time. Once having obtained the acceleration, they must then relate the force to it by using the second equation, which states that force = mass x acceleration. They can then find the force by solving the first equation for the acceleration and using the numerical result in the second equation:

$$a = \frac{30^2 - 0^2}{2 \times 45} = \frac{900}{90} = 10 \text{ meters/sec/sec}$$

That is, the acceleration, or change in velocity, is 10 meters per second each second.

$F = ma = 2 \times 10 = 20$ units of force (these units are Newtons).

Alternatively, the student can avoid solving for the acceleration, since it was not asked for, and combine the two equations to obtain a new equation relating the forces directly to the given quantities:

$$F = ma = m \left(\frac{v_2^2 - v_1^2}{2d} \right) = 2 \left(\frac{30^2 - 0^2}{2 \times 45} \right) = 20 \text{ Newtons}$$

The point is, it is not always one simple step from what is known to what is desired. In fact, even this example is a good deal simpler than some actual cases. Just as important to realize, however, is that no matter how many steps there are, they follow logically one from the other. You should therefore attack such problems not by trying to view the whole forest at the outset, but by blazing a trail from tree to tree. Only after you have completed the problem do you see the forest: that was the point of working the problem in the first place.

Having completed a word problem, you must then ask yourself: "Does my answer make sense? What is it telling me?" Students often relinquish their common sense, their intuitive ideas about nature, when they enter a science course, instead throwing their faith into "formulas." But these "formulas" are simply mathematical shorthand for what we know and what we can discover by experimentation. If an answer to a problem "looks funny" to you, go back and see if you made an arithmetic error. Then try to see if you have made an error in understanding the problem from the outset. This process of checking and rechecking your ideas is intrinsic to science; it is not a measure of how smart you are.

For quantitative sciences such as chemistry and physics, we cannot emphasize too strongly that the solution of word problems is the heart of the matter. The text material, the lectures, the homework: All are geared to teaching the student how to apply the concepts to actual physical situations as they appear in word problems. The student who says, "I understand the theory, but I can't work the problems" does not in fact understand the theory. "Understanding" a scientific theory means no more or less than being able to apply it to the solution of problems. They are the test of whether the student can reason formally, a prerequisite for really comprehending the nature of the intellectual enterprise we call science.

Let us say a few words here about biology. Although it tends to be less oriented to quantitative word problems than are chemistry and physics, it still requires the same kind of reasoning. Two articles of faith underlying all science are that Nature is logical and that human beings can understand her logic. In chemistry and physics this understanding is revealed through the solution of word problems; in biology it tends to be revealed in other ways, such as being able to describe the structure and function of various levels of organisms. But this description is itself logical, since Nature is logical. This description therefore requires the same sorts of reasoning skills that are used for solving quantitative problems. Thus biology homework is not all that different from chemistry and physics homework in the nature of its demands on the student. Biology is more than just "memorizing facts"; it is putting these facts together in a sensible way to understand the workings of Nature; it is "life science."

Equations and Graphs

For the science-anxious (as well as for the math-anxious) person, equations (formulas) and graphs are surrounded by a mystique that seems impenetrable. Yet, in reality, they are nothing more than a kind of shorthand for expressing scientific information. They need to be viewed in this light. As we have emphasized in our discussions of how to approach word problems, finding the "right formula," that is, the appropriate equation or equations, is the last step, after you have clearly delineated what you know and what you want to know. When you forget this, and fall back on your old memorization skills, looking for the right "formula" as if it will give you insight into the problem, then you tend to get confused and anxious.

Although it is not our task here to give a detailed explanation of how to work with equations that represent physical situations, we can, nevertheless, give some useful hints about using equations effectively. Let us suppose that you have followed the scheme we have outlined for solving word problems and have written down what you know and what you want. It is now time to choose the equation or equations that relate them. How do you make this choice? As you survey the possible candidates, and there may be more than one, you must first count the number of unknown quantities and ask, what do they represent? If more than one unknown quantity is present (i.e., does not appear in the column you have labeled What We Know), then either:

1. You need more than one equation; in fact the number of equations must equal the number of unknowns, or
2. You made an incorrect choice of equation.

The last problem we worked provides an example of both possibilities. We saw that there were two unknowns: acceleration and force. Therefore, we knew that we needed two equations. One was the simple one relating force to acceleration:

$$F = ma.$$

The second one related acceleration to initial and final velocities and to distance covered:

$$a = \frac{v_2^2 - v_1^2}{2d}$$

In choosing this one, as we said, we had to eliminate a number of others. These involved velocity and acceleration, but they also involved time. Since time is not asked for, nor is it given, we chose not to use any equation in which it appeared, opting instead for the one involving velocity and distance. If we *had* attempted to use an equation involving time, then it would have to be listed as a third unknown, along with force and acceleration. We would then have had to look for a third equation relating time to things we know, namely to velocity and distance. In fact, in this case, but not in every case, we would have found such an equation and been able to solve the problem, albeit in a somewhat longer fashion than we did. This approach illustrates another point about the solution of science problems: There is usually more than one way to skin a cat, and any logical approach that leads to the correct result is correct. Some approaches are simpler, or, as we scientists say, "more elegant" than others; they are all, however, equally correct. The notion that there is one and only one way to solve such problems is erroneous and leads to an irrational belief that only the person with the "scientific mind" can magically find the one true path while others cannot.

Another useful hint in using equations is to put in the actual numbers as late as possible. Carry out the algebra as far as you can, until the unknown is alone on one side of the equation and all the known quantities are on the other side, then put in the numbers. This was the second way we did the problem involving forces. Since we were not actually asked for the acceleration, we eliminated it algebraically, obtaining a final equation with force on one side and the known quantities -- mass, initial and final velocities, and distance -- on the other. Then, as the very last step, we put in the numbers and solved. This minimizes the chance of making an error in arithmetic. Although it is OK to solve for the acceleration as we did in the first method, if we had made a mistake in the acceleration, we would have got an incorrect number for the force, even though we had the correct equation relating one to the other. The fewer the "intermediate" quantities you calculate numerically, the less chance of making an error.

Finally, scientific quantities have units: meters, seconds, moles, kilograms, and so on. These units respond in the same way as numbers to mathematical operations. If you divide a distance by a time, you get a velocity. If the distance is given in meters, and the time in seconds, then the velocity is given in meters/second. It is essential to keep these units clear and to include them in your answer. The answer is meaningless if the units are not specified. Furthermore, keeping track of units can help you check if you have worked a problem correctly. If you are looking for a velocity, then after you have carried out your calculations, carry out the operations on the units. If the result comes out in meters/second, then your answer is at least "dimensionally correct," although the numbers might be wrong. If the units do not come out in meters/second, then you can be sure you have made a mistake and should rework the problem.

Let us now turn to the question of graphs and what role they play in science. As we said earlier, they are a shorthand for describing scientific results. In fact, they are even more: They summarize experimental data in such a way as to give the scientist strong clues as to the form of the theory that explains the experiments.

The first thing to keep in mind is that an experimental result always has some error associated with it; there is no such thing as the "perfect measurement." Therefore, when we use graphs to represent experimental results and we try to fit some smooth curve to our data points, the fit is never exact. The difference between an exact fit and the actual fit is an indication of the fact that mathematical theories are idealizations, "perfect" models of Nature's reality. Nevertheless, the nature of the curves that may be drawn through a set of data gives us clues as to the nature of the theory that describes the idealized model. Let us consider the following example:

Problem. Roll a little model cart down an inclined plane and measure how far it has gone in each interval of time, say, in each second (see Figure 3).

There are a number of ways in which we might make such a measurement. We could have the cart run along a wire to which a long strip of paper is attached and have the cart give off a spark every second, burning a little hole in the paper. We could then measure the positions of the holes to find out where the cart was. Or we could take a sequence of rapid photographs of the cart and compare them to see how far the cart had moved

Figure 3. Cart rolling down an incline.

in each photo. Let us suppose we have made the measurement and we list the position down the incline and the time at which this position was reached. We would get a table that looked like this:

Time (seconds)	Position (inches)
0.0 ± 0.05	0.0 ± 0.07
1.0	1.5
2.0	6.1
3.0	13.3
4.0	24.1
5.0	37.5

Notice that next to the first reading in each column is a number that indicates how much error we estimate is involved in simply reading the instrument (clock and ruler). Time is accurate only to within 0.05 seconds, and position to within 0.07 inches on either side (+ or -) of the actual reading that we make.

Now, since we are trying to find out how position varies with time, we can make a graph of the data. It is customary to put time along the horizontal axis and position along the vertical axis, but this is not necessary. Let us however follow tradition. Our graph will look like Figure 4. Now if we look at the graph, we see that we can draw a smooth curve through most of the points, and we do so (see Figure 5).

From what we know about math, we suspect from the curve that position may vary as time to some power: Call the position p and the time t, then $p = kt^2$ or $p = kt^3$ or $p = kt^n$ where n is any number, and k is a constant. We then can discover whether any of these is the case by a few simple tests. Let us start with the first possibility: $p = kt^2$. We can test this by drawing a new graph: of position versus the square of the time. The table that we draw up looks like this:

(Time)^2	Position
0.0	0.0
1.0	1.5
4.0	6.1
9.0	13.3
16.0	24.1
25.0	37.5

The position is plotted the same way, but along the horizontal axis we now plot t^2 rather than simply t. We obtain the points shown in Figure 6. Now these points look very much as if they fall on a straight line, so we draw one through them as in Figure 7.

83

We see that the line appears to pass through zero. We also see that position varies linearly with time squared, which was one of our hypotheses. The final step is to find the constant k in the equation $p = kt^2$. We measure this directly from the graph: it is the "slope" of the straight line, and we find its value to be 1.5, to within the error of our measurements. We now have a mathematical statement of the relationship between position and time as the cart rolls down the incline. How well does it fit the actual data? Let us make a table with three columns: time t, the calculated position $p = 1.5t^2$, and the actual positions that we measured:

Time t (Seconds)	Calculated Position	Measured Position
0.0	0.0	0.0
1.0	1.5	1.5
2.0	6.0	6.1
3.0	13.5	13.3
4.0	24.0	24.1
5.0	37.5	37.5

By comparing calculated and measured position, we see how much of our actual data differ from the idealized values predicted by the theory, and we get a measure of the accuracy of our experiment.

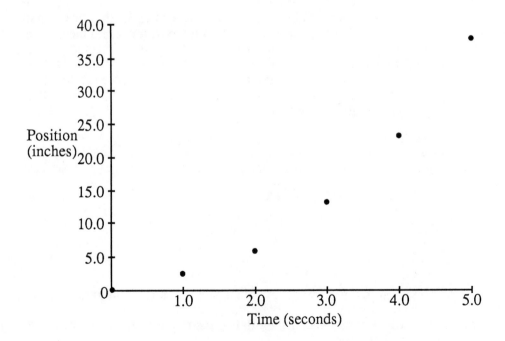

Figure 4. Graph of cart's position versus time.

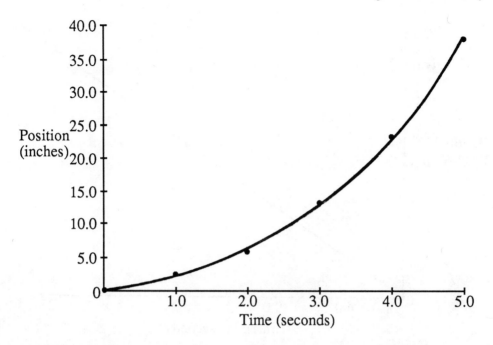

Figure 5. Graph of cart's position versus time, with curve drawn to best fit the data points.

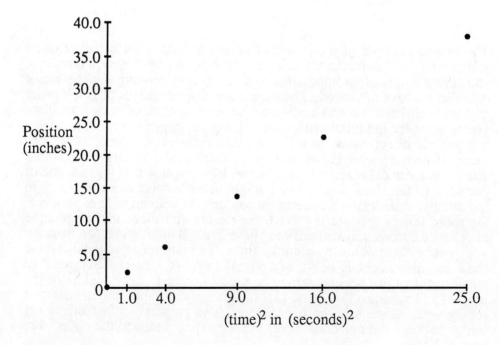

Figure 6. Graph of cart's position versus the square of the time.

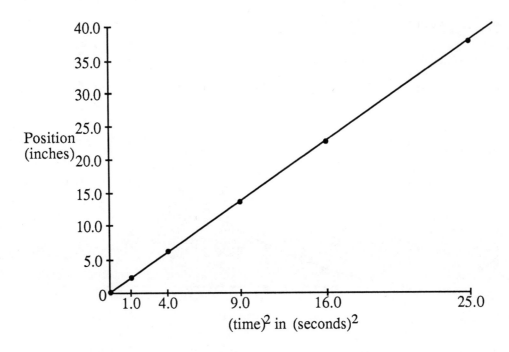

Figure 7. A straight line fit to the data points representing cart's position versus the square of the time.

This is one example of the power of graphs to help us visualize physical relationships and in fact to aid us in actually discovering these relationships, something that is often impossible to do by simply looking at the table of data that we have measured. There are many different kinds of graph paper that are useful in different situations. The purpose of all of them is to allow you to discover the relationships underlying the experimental data. They can provide direct visual information about things as disparate as rates of decay of atomic nuclei, growth rates of bacteria, orbits of stars around other stars, and chemical reactions. The power of a graph is that it grabs one of our senses: the visual sense, which might not otherwise come into play in our attempt to understand a scientific concept. As with any life experience, the more senses that are involved, the deeper and more meaningful the experience becomes. If you learn to play a musical instrument, for instance, you invoke three senses: auditory (how the instrument sounds), tactile (how the instrument is held), and visual (reading music). Science like music requires that several senses be used in order to get insight into the process. Graphs are the visual part of this sensory input, along with just seeing what happens in the experiment. The graph is in fact the visual bridge between the numerical results of the experiment and the underlying laws of Nature.

Taking a Science Exam

The science exam tends to be the situation in which most science anxiety comes out. Even if students are able to mask their anxieties during the course, they cannot do so under the pressure of taking an exam. In order to overcome this anxiety, you need to do two things: restructure your self-statements about your ability to pass such an exam and take the exam efficiently. We shall discuss some ways to do the latter in this section, and the former in the section on Negative Self-Statements.

First of all, the exam is an extension of the homework, so everything that we said in the section on working word problems holds for the exam as well. Focus on what is given and what is asked for. Only after you know this should you think about the possible equations that relate them. Check units. Leave the numerical work until you have simplified the algebra as much as possible. See what the answer is actually saying about Nature and see if it makes sense to you.

What kinds of additional hints do you need to take the exam? In the first place, do not necessarily work the exam questions in the order they are given. Look over the exam and work the ones that seem easiest first, then go on. Do not change your answers in the last few minutes, unless you are absolutely sure they are wrong. Most of the time you are not thinking that clearly at the end of the exam, so you will tend to change right answers to wrong ones.

If the exam has short-answer-type questions, such as multiple choice, and there is no penalty for a wrong answer as opposed to no answer at all, answer everything. Put something down for all the questions. Make a mark next to those you are unsure of, and return to them if you have time at the end.

The exam differs from homework in that the exam must test a wider range of material in a shorter period of time. You should therefore be looking for ways in which several different concepts appear in one problem.

Do not read more into a problem than is actually there. If one of your self-statements is "If I find it simple, I must be missing the point," then it is very tempting to look for more in the problem than actually is asked for. Be on your guard for this. Sometimes, of course, you will miss the point of a problem. But if this tends to be the way you view any problem that doesn't seem hard, you are probably reading too much in. The science-anxious student is often the one who answers a question that wasn't asked because he or she is distracted from what is actually on the exam by feelings of anxiety.

Prepare for the exam by getting a full night's sleep the night before. Avoid excess stimulants: They will make you more anxious. Avoid talking to your classmates before the exam. In fact, schedule things so that you arrive in the exam room no more than a minute or two before the exam is scheduled to begin. The more time you have to dwell on negative statements such as "I didn't study enough," or "Everyone knows more than me," the more anxious you will become. Don't give yourself this time.

When studying for an exam, do not try to memorize every possible problem in the text. Scientific reasoning is not simply the regurgitation of what you have already seen, it is the application of concepts to problems that you have not seen before. No matter how many problems a student memorizes, a science teacher can always think of a new one. This is one of the hallmarks of science: its infinite applicability.

While you are studying, try to avoid getting into the "forest-trees" dilemma. Focus on each section of your class notes before trying to put the whole thing together. This is often difficult to do, especially if you are thinking of how much more studying you still have to do to be ready. Nevertheless, it is essential that you stick to one concept at a time, because this is the only way you will learn it. If you try to get the "big picture" too early, you will simply waste your study time and frustrate yourself in the bargain.

Do not try to be perfect. You cannot memorize the whole course. There will be some sections that you will learn better than others. Accept this as natural. Trying to get 100 percent on every exam leads to anxiety and usually to a worse performance than if you had simply tried to do your best.

Avoid comparing yourself to your classmates. If you wish to do part of your studying with friends, by all means do so, but don't get into the game of "Did you study this?" Students express some of their anxieties by talking to others about how much or how little they studied. This game is never helpful because the players, intentionally or not, do not communicate any relevant or accurate information. In fact, the game is generally injurious, especially to the science-anxious student who is already prone to believe that everyone else has studied more.

In summary, much of the skill necessary for taking a science exam can be expressed in a few words: "Focus on the task and avoid extraneous anxiety producing situations."

Laboratories

The purpose of the science laboratory is to give students hands-on, concrete experience with the phenomena that are being discussed in the lecture. The way this purpose is achieved varies from school to school, as does the degree of success. Science teachers are constantly experimenting with new forms of laboratory instruction. It is very rare indeed that a science department finds its laboratory courses entirely satisfactory. This may be due to many factors, among them, the availability of funds for the necessary equipment, the different levels of concrete experience the students bring to the course, the difficulty in keeping their lab and the lecture in phase with each other.

Lab courses range from the so-called "cookbook" lab, where every step of the experimental procedure is spelled out and the students are graded objectively on how well they complete each step, all the way to the "open-ended" lab, where students are placed in a room with equipment and asked to plan and carry out experiments to measure certain properties or test certain laws of Nature, in which case the grading procedure is more subjective and more difficult. As you can see from considering these two extreme types of laboratory course, and recognizing all the possible variations in between, the lab has a number of purposes: to give the students experience with equipment, to teach them to be systematic, to put them face to face with Nature without the intervention of theory, to measure their creativity in testing Nature's laws, and to relate their lecture experience to reality.

What can you do to get the most out of the lab experience? The first thing is to be prepared beforehand. This means carefully studying the experiment and thinking about what the likely outcome of the measurements will be before you ever come into the lab. If you have difficulty visualizing the setup just from reading the lab book, try to get access to the lab to look things over before your assigned lab period. Prepare all data tables beforehand. The less writing you have to do in the lab, the more time you have to focus on what's going on.

Once in the laboratory take an active part. Students who are anxious about the laboratory part of science courses tend to hold back, to let their lab partner handle the equipment while they just write down the numbers. This is like watching tennis. You can't learn to play unless you play. Attack each piece of apparatus separately and try to understand what it alone does, just as you attack each part of your lecture notes separately. Then try to understand how the whole setup works.

Take repeated sets of data, and do not expect them all to turn out exactly the same. Scientific laws are idealizations; all experiments have error associated with them. Repeated readings allow you not only to see if you have made actual mistakes, but also give you an estimate of the intrinsic random errors associated with the measurement itself. Do not throw away bad data points unless you can see where you might actually have made a mistake in the measurement. Look for sources of error in the limitations of the lab experiment rather than in some negative self-statement about your experimental ability.

Stop at various points in the laboratory period and see if what you are doing makes sense. Ask the same questions that you ask about word problems: "What am I given? What am I trying to find? What relates the one to the other?" Make a few rough calculations along the way, to see if your data makes physical sense. Are the numbers more or less what you expected? If not, why not? Were your expectations incorrect or did you make some systematic error in the measurement or does the experiment not measure the property it is supposed to measure to the precision that you anticipated?

If you keep these very rational questions in mind as you carry out the lab experiment, you will not have time to think of a host of irrational self-statements that have nothing to do with the laboratory, and you will probably carry out the experiment with an efficiency that will surprise you.

Chapter 4

Calculators

Before we close this section on science skills, we must say a word about the use of calculators in science and enumerate their advantages and disadvantages.

The advantages are fairly obvious. Calculations that take a very long time when done by hand can be done rapidly and with great accuracy on the calculator. The simple operations such as multiplication and long division and the more complicated ones such as finding trigonometric functions, exponentials, logarithms, or square roots now take less than a second on a hand calculator. Therefore the student is free to focus on the science rather than on the math that is its tool. More problems can be worked out in less time. Arithmetic too long and complicated for the hand calculator can be programmed on computers, which are available at most universities and many high schools. So much of the mathematical drudgery that used to accompany science has been taken over by machines.

The disadvantage is that many students have become too dependent on these machines. They are increasingly less familiar with mathematical operations per se and more dependent on the calculator to give them even simple answers. There are few experiences more frightening to a science teacher than to see a student find the sine of 90 degrees by punching it out on a calculator and getting the answer: 1. Students lose sight of the laws --mathematical and scientific -- underlying the development and the importance of such mathematical functions. For a math teacher to try to explain "logarithmic dependence" to a group of students whose only experience with logarithms is the calculator button called LOG is frustrating indeed.

For the scientist the situation is equally frustrating, but in other ways. Science is quantitative. It asks questions such as "how much?" and "how fast?" about natural phenomena. In order to do science, you must be able to make estimates about the size of such things as well as the accuracy to which they can be measured. Using a calculator allows the student to get away without having to think about these estimates. The answer on the calculator display becomes gospel, with no thought as to whether it is meaningful or not. If you accidentally push the wrong button, you get the wrong answer, but how do you know if you have no experience with estimating? Even if you push the right button, the answer is usually given to seven or eight figures on the calculator. How many of these are meaningful? If the numbers you put in were only good to three digits, then the last five in the calculator's answer are nonsense and you must throw them away.

What is all this diatribe leading up to? Just this: The skill needed to use a calculator effectively in science is the skill of knowing when not to use it. After you have worked a problem on the calculator rework it by hand, not in detail, but with all the numbers rounded off, just to see if your answer makes sense. This was common practice in the old slide-rule days, since that device never told you where to put the decimal point; therefore, you had to do a so-called "order-of-magnitude calculation" to find whether an answer was, say, 7.43 or 74.3 or 743. The calculator gives you the decimal, but you must still be able to check that it is in the right place, i.e., that you have not made an error. Otherwise you become the slave of the calculator: Without it, you have no idea what the answer should be.

Now making estimates is not a new skill. Almost everyone has to do it at some time or other: figuring out how much money to bring to the supermarket, for example. The problem is that students with calculators tend to let this skill atrophy just when they need it most: when they are trying to study science. In homework problems, on exams, in laboratory calculations, by all means use the calculator, but by all means check your answers by doing rough "order-of-magnitude" hand computations. This is what all professional scientists do -- because without it they would lose sight of what they are doing and would reduce science to a list of numbers.

This completes our discussion of some of the skills that you need to do science and some hints about how you can develop them. Now let us move on to the second phase of the attack on science anxiety: how you can discover some of your personal anxiety producing self-statements and how you can begin to overcome them.

Negative Self-Statements: How to Cope with Them

A good deal of the fight against science anxiety involves "cognitive restructuring"[2]: recognizing when you are feeling anxious; tuning in to the things you are telling yourself, the "negative self-statements" that cause the anxiety; examining these statements for the irrational beliefs behind them; and finally, producing "coping statements" -- new self-statements that counter your anxiety and allow you to function better in science-related situations. As you read this section, write down the situations in which you feel anxious. Try to find your negative self-statements among those listed below; or, even better, try to see what your statements are without looking at the list beforehand. Then see if you can generate coping statements with the help of the discussion that follows. Naturally, one cannot think of every possible set of negative (or coping) statements, but it is hoped that you will find some in the list that are related to your own. We have found that science-anxious students share many of the same negative self-statements, so it is likely that if you are science anxious you will find things here that will help.

The first task is to recognize when you are anxious. Sometimes it is quite obvious: sweaty palms, upset stomach, or other physical symptoms. Sometimes the signs are a little subtler. You may simply feel distracted while you are trying to study. Or you may discover that you are tapping your foot or chewing your fingernails, although you don't feel any overt anxiety. Your mind may go blank on an exam or during a lecture. You might suddenly get a headache. All these can be signs that you are feeling anxious: Different people express their tensions in different ways. Whatever your form of anxiety, the first thing you must do is recognize it

91

for what it is. Be on the lookout for clues that you are feeling anxious. Some of the symptoms are psychological, such as tension and nervousness; some are physical, such as stomach aches or even rashes.

As you become aware that you are feeling anxious, write down the situations in which the feeling of anxiety occurs. Be very specific. Do not simply say, "I feel anxious when I am studying chemistry." Say, "I feel my stomach muscles tighten while I am in my room looking at a homework problem in chemistry." In order to combat your anxiety, you must be able to visualize clearly the conditions under which it occurs.

Once you have made a list of when you feel anxious and how you experience your anxiety, you will begin to look for the connection between the stimulus, that is, the event that leads to your anxiety, and the consequence, that is, the anxiety itself. The connection is some negative self-statement, something you tell yourself, and of which you are not usually aware, that turns a neutral event such as studying science into a source of anxiety. In the Clinic we refer to this procedure as the A-B-C sequence: the event A led to the anxiety C because of the intervening self-statement B. The task now becomes the identification of the self statement or statements. Since these are not usually conscious statements, you need to tune into them, to listen with a "third ear": the one facing inward. In the Science Anxiety Clinic, the psychologist and the scientist elicit from the students their negative self-statements over a period of a few weeks. Since this is a chapter on self-help, the best we can do is to give a few hints about how to listen with the "third ear" and then list some of the common self-statements that we have elicited from students in the Clinic.

The time to tune in to your negative self-statements is *when you are feeling anxious*. It is very difficult to sit back in a calm moment and try to figure out what you tell yourself when you are not calm. But once you have learned to recognize when you feel anxious, you must stop and think during the experience and write down what comes to mind, even if it seems silly or odd. Our negative self-statements are odd. They introduce irrationality into our daily attempts to function in school and elsewhere. It may be difficult at first for you to own up to these statements. You may feel somewhat embarrassed. Plunge ahead anyway. The only way to counter negative self-statements is to face them head on. Keep in mind that you are not weird or crazy. Everyone who has science anxiety, in fact, everyone who feels a lot of anxiety, has negative self-statements somewhere generating it.

Here is a list of some of the statements that students have made in our Clinic:

1. No matter how much I study I'll never understand science.
2. If I don't get a perfect score, it means I didn't study enough.
3. If I'm not studying all the time, I'm not studying enough.
4. I should have studied more.
5. I probably studied all the wrong material.
6. I'm not worried enough about this science exam: I'll probably do poorly.

7. If this problem looks easy, it's probably because I don't really understand what is asked for.
8. I'll never get this lab experiment to work.
9. I could get an A in this science course if only I worked hard enough.
10. There's just too much material to study here. I'm overwhelmed.
11. I should know this material.
12. If I can't understand this science material, I shouldn't even be in school.
13. All this trouble I have grasping science just proves how incompetent I am.
14. If I don't do well in science, I'm worthless.
15. If I ask the teacher a question, I'll show how dumb I am.
16. What will the teacher think of me if I can't answer this question?
17. Other students ask such intelligent questions. Why can't I?
18. Other people seem to know the secret of understanding science. Why don't I?
19. Everyone around me is writing answers to this exam. I can't think of a thing.
20. Everyone else understands the material.
21. I'm not really as smart as the other students.
22. Everyone else can do it except me.
23. Science is too hard for normal people.
24. Only superbrains can understand science.
25. Boys are supposed to be good in science.
26. Science is not for girls.
27. Science is not for me.
28. I'm scared of science, and I'm all alone in my fear.
29. I can't be expected to know this.
30. I don't have a scientific mind.
31. I don't really have to know any science to be successful in my chosen career.
32. If I miss this problem on the exam, I may fail it. Then what if I also fail the next exam? And the next?
33. If I don't get an A in science, I won't get into medical school (dental school, nursing school, graduate school, college, etc.).
34. My future hinges on this science course.
35. My folks are counting on me to do well.
36. I should be studying, but I want to play.

Let us see what characteristics these statements exhibit that make them negative and anxiety producing. First of all, they are filled with loaded words, words that take science learning and turn it into an emotionally trying event. Words such as "should," "never," "expected," "worthless," "incompetent," and "perfect" occur throughout. These words distract you from focusing on the task of comprehending science because they take you out of the here-and-now and put you into some hypothetical threatening future situation, such as an exam or some other place where you will be "shown up" for the incompetent you are.

Second, the statements introduce a measure of irrationality into an otherwise rational situation, namely, studying science. The irrationality is introduced in any number of ways. Statements 1 through 10 might be described as examples of abdicating your sense of judgment. Instead of recognizing the rational aspect of studying science -- that it takes work, but is not magical and that you can decide when you have studied enough -- you ascribe all the power to some vague external force that will "never" allow you to grasp science. You see yourself as powerless: powerless to decide when to stop studying, powerless to make an experiment work, powerless to know what to study, powerless to decide when a question is easy or hard. So you revert to a primitive mode of trying to affect the outcome of an event perceived to be out of your control; namely, praying to an idol. The modern form of this is called worrying, as in statement 6. Not the ordinary, reasonable sort of worrying, but a kind of institutionalized worrying, as if enough worry will improve your chances to succeed in science. Statement 10 represents the science-anxious person who simply gives up. Rather than focusing on science study as a series of separate tasks taken one at a time, the anxious person sees only the forest, not the trees, only the whole mass of material, which of course appears overwhelming.

Statements 11 through 16 are irrational in a different way. They hook your performance in science directly to your sense of personal worth. Each difficult task in the study of science is taken as a sign of your basic worthlessness. Interestingly, among science-anxious students each success in a science course is ascribed to luck. So it is easy to see how irrational these statements are. Just as statements 1 through 10 are examples of abdicating the power to make rational judgments, statements 11 through 16 are examples of abdicating the sense of self, tying it entirely to performance in the study of science.

Statements 17 through 28 might be titled "making comparisons with a mythical norm." In these statements you compare yourself with others, doing so on the basis of little information or misinformation. In statements 17 through 22, you see others as normal, as able to understand science, as relaxed and intelligent, while you are abnormal. These kinds of negative self-statements are highly irrational, because actually the science-anxious person has little evidence that everyone else is smarter. In fact, the word "everyone," like "never," is a tipoff that the statement is irrational. How, for example, do you decide, as in statement 17, that other students' questions are intelligent? Either you do not understand the question, in which case you cannot assess whether it is intelligent, or you do understand the question, in which case the questioner is not any smarter than you! In fact, statement 17 is really saying what statement 7 said: "If I understand it, it can't be very profound, and if I don't understand it, then it is probably profound."

Statements 23 through 27 are the mirror images of 17 through 22. In these latter statements, the people who can understand science are the "abnormal ones," while you are in the "normal" group, i.e., the group that cannot understand science. (Statement 25 is of this sort when expressed by females. When expressed by males, it is in the same category as statements 17-22.) In either case, you define yourself as intrinsically in the wrong group, with no hope of passing into the other. Of course in statement 23 through 27 there is the consolation that you are not alone and are "normal." But it's not much consolation.

Statement 28, the last of the "mythical comparison" statements, is a very popular one with science-anxious people. It completely and utterly isolates you from the rest of the world. Only you have a fear of science, and no one else can understand your problem. As you might imagine, this world view produces enormous anxiety. (One great virtue of the Science Anxiety Clinic is that it can quickly undercut this irrational statement, because the room is filled with science-anxious people.)

Statements 29 through 36 might be categorized as "adding to the pressure." The first three are the most deceptive because they look like statements to relieve pressure, to minimize the importance of learning science. If they were made in an emotionally neutral setting, then they might be accurate, albeit unfortunate, assessments of the situation. For example, an art student looking to an advanced biology textbook could accurately say, "I can't be expected to know this," or "I don't need to know science for my career." Even the statement "I don't have a scientific mind," although misguided as we saw earlier, need not produce anxiety when uttered by this hypothetical art student. However, these statements are usually voiced in a very different context. The student who is required to take a science course or the adult who encounters something technical that is job related -- these are the people who seek refuge in those statements. But there is no refuge to be found. You are expected to know the material, it does affect your career, and the claim that you don't have a "scientific mind" is irrelevant.

Statements 32 through 36 lay the pressure on in a much more direct way. The first three are prime examples of what psychologists call "catastrophizing": extrapolating from a limited situation, such as a science exam, to an irrational and terrifying set of consequences. The panic evoked by this sort of negative self-statement tends to paralyze the person at just the time when efficient performance in science is essential. The last two statements are less connected to any particular science event, such as an exam. On the other hand, they tend always to be there in the background of your thoughts when you are trying to accomplish any science task: reading a textbook, working problems, or just studying science material. They therefore provide a continual level of distraction, which makes it harder to concentrate.

Of course this is not an exhaustive list of negative self-statements.

Nevertheless, the common types of statements have been well represented. We should also note that not all the statements appear to be science related, and, in fact, some students generate these statements about other subjects (I am not talking just about math, which generates anxieties similar to those experienced about science, but about, for example, foreign languages). However, the prevalence and intensity of these statements appear to be much greater in people facing the prospect of having to learn science.

Once you have identified your negative self-statements, you must begin the search for the "antidote," the so called "coping statement." This is not simply the opposite of the negative statement. It does no good, for example, to tell the person who is scared of science (statement 28) that "There's nothing to be afraid of." This simply discounts the fear and replaces it with the idea that the anxious person is a little crazy. Other sorts of positive-sounding statements, such as "You can do it if you try" or "Don't worry about grades, just do your best" may be well meaning, but they are useless. They are, in fact, merely a form of cheerleading that does not address the fear itself.

The basic requirement of the coping statement is that it focus on what is rational and what is irrational about the negative statement and separate the one from the other. Let us take negative statement 1 as an example: "No matter how much I study, I'll never understand science." What is rational? That it takes serious study to understand science. What is irrational? The mystical prediction of the future embodied in the words "no matter how much" and "never." A good way to approach an appropriate coping statement is, therefore, to question the irrational assumptions:

" 'Never' is an absolute and final word. How can you predict the future? On what evidence do you base the assumption that no matter what you do it will prove fruitless? How can you be so certain of failure?" Thus we see that the pessimism embodied in the negative self-statement is not countered by some "cockeyed optimist" but rather by an "objective observer" who can focus on the irrational beliefs underlying the negative self-statement and attack these beliefs with reason.

I must emphasize at this point: It is not sufficient to "think through" science anxiety. People do not necessarily change the way they feel and act because they understand something intellectually. Hence, even after you have found coping statements (no small task in itself), it takes time for these new messages to integrate themselves into your mental frame of reference and replace the negative statements. (This is one of the reasons that we use a multipronged approach in the Science Anxiety Clinic: While the students are integrating coping statements over a period of weeks, they are also learning relaxation techniques and new science skills.)

Coping statements, like negative statements, vary from person to person, so it would not be particularly useful to run down our list of negative statements and provide a coping statement for each one. In fact, the most useful thing for the science-anxious reader to do is to try to develop personal coping statements. We can, however, provide some general guidelines. The first has already been stated: Find the rational core and the irrational trimmings of the negative statement. Then counter the irrational part with objective statements, rather than with "cheerleading" statements. The next thing to do is to change your semantics, i.e., the way you tend to phrase things. Look for statements involving "should" and "must" as well as generalizations such as "never" or "always" or "everyone but me." Start to eliminate them actively from your vocabulary: They are not rational. Begin focusing on the here-and-now. A good, objective coping statement useful for almost any negative statement is, "This negative statement is distracting me from the task at hand. It does not help in learning science. Focus on the task."

For the first 10 negative statements, the ones we call "abdicating your sense of judgment," coping statements naturally take the direction of recapturing this sense. "No amount of studying guarantees perfection; some amount of studying is certainly enough; worrying guarantees nothing at all; if I have studied, then some problems will look easy." These are the sorts of directions that coping statements can take to help you recapture your sense of judgment. Sometimes a textbook or lecture seems confusing because it is confusing: poorly written, poorly presented, or even wrong. Entertain this possibility as you attempt to study science, in the same way you do when you study other subjects.

Statements 11 through 16 must be countered by coping statements that separate your self worth from your performance: "What does it mean to be 'worthless'? Isn't this an irrational word? Isn't the teacher there to answer my questions? Isn't asking questions a good way to learn?" And suppose the teacher is unresponsive, or even, as unfortunately happens from time to time, sarcastic? Whose problem is that? Yours or the teacher's? You have done nothing wrong by trying to understand science.

Statements 17 through 28, the comparisons with the mythical norm, are best countered by questioning the basis on which they are made: "What evidence do I have that the other person's questions are clever and mine are not? Isn't it rather magical to think that everyone but me knows the secret of understanding science? Who says that only certain people -- superbrains, or boys -- are the only ones who can understand science? What is the evidence to support this claim?" (There is none.)

The catastrophizing statements can be dealt with by fantasizing the worst possible consequences of, say, doing poorly on an exam and allowing the fantasies to become as grandiose as possible -- and then recognizing how irrational they are: "One event does not shape my future forever after. What is the most awful thing that will happen if I can't work this chemistry problem? Is that realistic? How terrible are the actual consequences?" These sorts of questions make up the kinds of coping statements that counteract the catastrophic negative self-statements that produce science anxiety in highly charged situations such as exams.

It is sometimes the case that once you have been able to counter and neutralize some negative self-statements others begin to surface. Don't be dismayed. This is quite natural: Getting in touch with the things we tell ourselves is not a one-shot deal. It is more like peeling an onion, with successive layers appearing after we have dealt with the outer ones. So, if one set of negative statements gives way to another, keep using the same techniques, developing new coping statements. Be patient. Nothing happens overnight. But the techniques outlined here have one great advantage: They work. And they have worked on many science-anxious people. They will very likely work for you. To believe otherwise is itself a negative self-statement.

Relaxation and Desensitization

The third and final technique for combatting science anxiety is called "relaxation and desensitization training." As the name suggests, it is actually a combination of two components:

1. Relaxation: training in relaxation of various muscles on command.
2. Systematic desensitization: training in remaining relaxed even in the presence of stimuli (such as science exams) which previously caused you anxiety.

This technique was developed a quarter of a century ago by the psychotherapist J. Wolpe[3], for working with anxious patients. The principle behind it is called "reciprocal inhibition" which simply means that you can't feel contradictory sensations at the same time; one inhibits the other. In particular, you can't be both anxious and relaxed. Thus, if there were a way to train people to respond to science stimuli by *relaxing*, their anxiety would be inhibited. This is precisely what we do in the Science Anxiety Clinic (see Chapter 6). We first use a technique developed in 1938 by E. Jacobson:[4] "deep muscle relaxation." Students are taught to relax various groups of muscles on command, until their entire bodies are relaxed. There are standard tape cassettes which provide these commands,[5] and you will need to purchase one in order to carry out this program yourself. After a few weeks of listening and responding to the tape, you will probably be able to relax by giving yourself the same commands.

During this time, you must develop your own "science anxiety hierarchy": a list of science-related stimuli, in order of least-to-most anxiety provoking. A sample hierarchy appears below. Once you have developed your own, and at the same time you have learned deep muscle relaxation, you are ready to begin the desensitization procedure. Since you are not in a Science Anxiety Clinic, you will need to find a relative or friend who is willing to work with you. You must first get yourself into the relaxed state. In addition to the relaxation training, you may also want to picture yourself in a peaceful scene -- at a beach, or in a meadow, for example. View the scene from the *inside*, not from the outside looking in.

Now your relative or friend reads to you the lowest anxiety producing item on your hierarchy, and waits about 10 seconds. If you feel anxious, lift your index finger. He or she then repeats the procedure. Use the peaceful imagery and test that your muscles are relaxed, concentrating on specific areas of tension. Eventually, you will be relaxed during the reading of the stimulus and the 10 seconds following. It is then read again, this time followed by a 3-second delay. If you remain relaxed, he or she goes on to the next item. Some may, surprisingly, evoke no anxiety, even though you put them in your hierarchy. If any subsequent stimulus provokes anxiety, your helper must drop back one item, and then come up again.

Chapter 6 describes in more detail this procedure as we employ it in our Science Anxiety Clinic.

Sample Anxiety Hierarchy for a Physics Course

1. On registration day, you are signing up for physics.

2. On the first day of classes you are standing in the bookstore thumbing through the textbooks for your physics course.

3. On the first day of classes your physics instructor has presented an outline of what is to be covered during the semester.

4. One month after classes have started you are sitting in your physics class. As the instructor invites the class to participate and ask questions, you raise your hand to ask a question. The instructor is explaining one of the homework problems that seemed to confuse most of the class.

5. You are studying the chapter covered in your last physics class. You have turned to the end of the chapter to do the assigned problems. You are sitting and reading the page of problems.

6. You begin working on the first problems in the chapter which you have just read in your science text. It is the same material your instructor covered in the last class.

7. While trying to solve one of the physics problems in your homework, you use several different formulas and get several different answers.

8. You are studying for the weekly quiz. You have read the material once and think to yourself that you will review it once more before the quiz.

9. While you are taking the quiz, your instructor walks past you and pauses, looking at your paper.

10. During a lecture in your physics course, you don't understand what your instructor is saying, but you notice that everyone around you is nodding their heads in understanding. They are rapidly writing down notes.

---

11. In the lab you are working by yourself trying to set up an awkward piece of apparatus. You notice the lab instructor is standing behind you, watching what you are doing.

12. You are studying the night before the final exam.

13. You have just been handed the final exam.

[1]S. Weinberg, *The First Three Minutes*, New York: Basic Books, P. vii, 1977.

[2]A. Ellis, *Reason and Emotion in Psychotherapy*, New York: Lyle Stuart Press, 1962.

[3]J. Wolpe, *Psychotherapy by Reciprocal Inhibition*, Stanford: Stanford University Press, 1958.

[4]E. Jacobson, *Progressive Relaxation*, Chicago: University of Chicago Press, 1938.

[5]For example -- T.H. Budzynski, "Relaxation Training Program" (1974), available from Biomonitoring Applications, Inc., New York.

5

Fighting Science Anxiety: The Role Of Teachers And Parents

The previous chapter has attempted to help the science-anxious person deal with his or her problem. However, we have seen that the source of this problem is external: societal myths and pressures, the way science is taught, the availability of appropriate role models, and so forth. This chapter addresses some of these issues; in particular, what constitutes good anxiety-free science teaching, what parents can do to help their children approach science without anxiety, and what the connections are between science anxiety and other personal problems.

Science Teachers

The student's primary contact with the ideas of science is through the science teacher. This is more true of science than, say, of English. An intelligent young person will probably read, whether or not every English teacher he or she comes in contact with is a good one. An adult's reading habits are likewise probably fairly independent of the quality of past English teachers, although, obviously, the better the teachers, the more likely the recipient of the teaching will be to delve more deeply into the subject. However, it is probably accurate to say that a bad English teacher will not produce a person who does not read at all. A bad science teacher, on the other hand, will almost certainly produce a person who avoids any and all contact with science forevermore. It is therefore essential that we take a close look at what constitutes good and bad science teaching. If we can change some of the ways in which science teaching results in science anxiety, we will be striking right at the heart of the problem.

Chapter 5

Teaching Techniques

Science is the exploration of nature. Science teaching must communicate that sense of exploration and must bring the student along as a partner on the journey, not simply as a passive observer. The student's faculties must be engaged to the utmost in the enterprise; otherwise he or she learns a little about science but does not learn how to do science. Without this sense of being able to do science, the student remains scientifically illiterate and, in most cases, frightened of science. Therefore the good science teacher must find ways to involve the student.

The first way, and the one most observed in the breach, is to limit class size. The more students in a class, the more distant they are from the teacher, the less likely they are to become engaged in the adventure of science. To my mind, classes of more than 60 students are at cross-purposes with the goal of teaching science. The teacher becomes a distant, forbidding figure up at the lectern with a microphone; there is little or no time for questions and answers, thus depriving the students of the opportunity to question and the teacher of the feedback as to what is getting across. High schools still have fairly stringent limits on class size, usually less than 40. In many large universities, however, classes are as large as 500.

I think it is imperative that this trend toward large classes in the sciences be reversed. This of course means hiring more faculty, not an easy task in a period of declining enrollments and scarce financial resources. Perhaps the acuteness of the problem of science anxiety and the concomitant science illiteracy among our population will be recognized by the people who make educational funding policy, and more money for faculty will be made available. This will not happen by magic. As long as students are required to take science courses (or other courses) to fulfill degree or professional school requirements, and as long as they tolerate large classes, then large classes will continue. If students, and their parents, who are usually footing the bill for schooling, make known to school administrators that large classes are unacceptable, then, given enough pressure, the administrators will try to solve the problem. In fact, I think that a large volume of complaints provides good ammunition for school administrators to convince trustees and government funding agencies of the need for additional faculty to reduce science class size.

What holds for science lecture classes holds even more for laboratories. This is the place where students are supposed to get firsthand experience with doing experiments. Unless lab class sizes are severely limited, namely, no more than two students for each experimental setup, the lab is useless. I know of schools where labs have been virtually abolished; students simply walk from demonstration to demonstration, taking notes on what they observe. This is ridiculous! If enrollments in a science course grow so large that labs turn into demonstration sessions like this, then the entire science course has to be revamped. Students approach the science lecture and the science laboratory with fears and misconceptions, thanks to years of social conditioning. The very least the science teachers can do is to undercut some of these fears by making the lecture and lab of reasonable size and therefore accessible.

There are other ways in which the teacher can communicate that science is exploration rather than a cut-and-dried summary of facts. One is to mix the standard "linear" approach to science teaching with the "mystery story" approach. The linear approach is the one most of us are used to: the logical, step-by-step exposition of a concept, complete with proof of each step on the blackboard. This is an important and useful approach because it reveals the internal logical consistency of science. However, it tends not to engage the students in the enterprise. The "mystery story" approach to science teaching involves presenting the "evidence," for example a series of demonstrations, and then looking for a logical explanation for all the phenomena. The explanations will not of course come all at once, but will be developed in the course of the lectures--this is where the linear approach is useful. But constant reference to the demonstrations will keep the students in contact with the sense of excitement surrounding the discovery of scientific laws that explain a host of hitherto mysterious phenomena.

There is another reason why the science teacher should avoid exclusive use of the linear approach: It gives a false view of how science is actually done. In particular, when students are assigned science problems for homework, the way to approach them is not the same as the way the teacher works them out in class. However, it is the way the teacher and the professional scientist approach problem solving. To put things in terms of our discussion in earlier chapters, science problems are solved by a combination of convergent and divergent thinking, but they are usually presented as if they had been solved entirely by convergent thinking. A scientist attacking a new research problem does not immediately see one and only one route to the solution. Nor does a student attacking a homework problem. But the science teacher puts the problem solution on the blackboard the way the scientist writes a paper about the research problem after solving it: linearly, logically, devoid of the context of random thoughts, false starts, and blind alleys that accompanied the process of solution. This gives students a deceptive view of how science is actually done and feeds into their negative self-statements that they don't have scientific minds, that they are not smart, and so forth. It is worthwhile for science teachers to discuss the process that went into their solving homework problems, not this time around, but the first time, when they themselves were students. It is even useful to explore one or two of the blind alleys that students might have taken in the course of trying to work the problems. This all takes a bit of time, but it helps demystify the process of doing science, thereby lowering anxieties as well as sharpening problem-solving skills.

The science teacher also needs to be very careful in the use of audiovisual aids. Pictures of various scientific phenomena are fine, but the use of slides and transparencies to present material that could be presented on the blackboard is fraught with danger. The purpose of the blackboard is not only to present material in a dynamic way (something a transparency does not do) but also to slow the teacher down. Students need time to take notes and to absorb a little bit of the ideas presented: using slides and transparencies for "speed and efficiency" has just the opposite effect. Furthermore, whenever the lights are dimmed in a lecture room the students become passive observers, like a movie audience. There is no longer eye contact with the instructor and therefore no way the instructor can monitor whether material is getting across. Science teaching needs all this feedback if it is to be anxiety free.

One of the other lessons of science which a good teacher can communicate is that most often, good science is a group enterprise. There is something about the back-and-forth playing with ideas that leads to inspiration. The same group process can be applied to science courses. Students should be encouraged to study together, to ask and answer each others' questions, and to compare their various interpretations of a scientific concept. There is no better way to learn a scientific idea than to have to explain it to someone else. Furthermore, students are often more open to hearing the explanation from a peer, whom they expect they can understand, than from a teacher, whom they may not expect to understand. They are more willing to question each other than even the most sympathetic of teachers. Since science seems to lend itself to a group approach, there is every reason to instill this approach early, with the first science course a student takes.

A science teacher has another task, one related to the notions of convergent and divergent thinking. He or she must teach by example how to separate relevant from irrelevant information in understanding a scientific idea. This is no simple task. Let us take as an example a laboratory experiment in physics, one in fact that we discussed in connection with the Piaget model of learning. Suppose a student is presented with a simple pendulum consisting of a ball suspended by a string from some fixed mounting (see Figure 2). The student, let us say, has a choice of balls of different weights and different materials and strings of different lengths. The pendulum is set in motion by pulling the ball sideways a certain distance and letting go--and the student also has the freedom to choose this distance. There are many variables here. The teacher can choose one of two extreme ways to present the demonstration, or somewhere in between. The "minimum clue" extreme is to ask the student: "Find what variables affect the period of the pendulum; that is, the time it takes to make one complete pass and return to its original position." The "maximum clue" extreme is to ask: "Show that only the length of the string affects the period of the pendulum." In both cases, the student has to carry out a series of measurements using different balls, different strings, and different initial distances. In the first case, however, the student is given no clues and must decide what is relevant to the pendulum's period and what is not. In the second case, the student is in effect given the answer, and the experiment really consists of showing that the answer is right.

Another example is the science test in which some irrelevant information is provided and the student must decide what is not needed to solve the problem. This contrasts with the test in which only that information that is needed is given. Now students bring to these tasks different abilities and experiences to which the teacher must be sensitive. Some students are "field-dependent";[1] that is, they depend heavily on clues and must expend a lot of energy trying to separate useful from misleading clues. Other students are fairly "field-independent": they seem able with relative ease to focus in on just those clues that will help them solve the problem. In any case, there will be both types of students and every type in between. Since actual science involves many misleading clues, it is important that the science teacher spend some time making students aware that part of their task is to decide what clues are irrelevant. Otherwise, science will be once again perceived as an example of convergent thinking: just enough clues to give the student one path from question to answer. This is not to say that every question in a science course should contain irrelevant information. Those problems that are used to introduce a new concept should contain only useful information. Once the concept is familiar, however, the student needs practice in applying it to a host of realistic situations. Here the introduction of some irrelevant information would seem to be in order.

Just as the teacher has to walk a careful line between the field-dependent and the field-independent student, so must he or she strike a balance between the presentation of material in a formal versus a concrete way. As we discussed in Chapter 3, about half the students who enter college science courses are formal-operational thinkers, while the others are still concrete-operational. The laboratory experience certainly caters to the concrete, providing hands-on experience with the phenomena discussed in the lecture. The lecture itself, however, must also mix concrete with formal. This can be done in the same way (and in fact simultaneously) with the "mystery story" approach to science. Demonstrations of phenomena in class are concrete; their explanation is formal. Thus the science teacher can, by a judicious mix of demonstration with logical exposition, reach both the concrete-and-formal-operational students.

One final ingredient: The science class must be a place where questions can be asked and answered. The tension between the need for questions from students and the need to lecture according to a given syllabus within the time constraint of a semester or a quarter is well known. Most science courses at colleges schedule at least one hour a week to devote to questions: This is the so-called "quiz-session" or "recitation session." Here students should be able to ask the instructor whatever was not clear to them in the lectures. Here also, they work science problems.

While the quiz session is extremely useful, it is sometimes not enough. This is particularly true when the quiz session instructor is not the same person as the lecturer. One important function of student questions is to give the lecturer feedback about what the students do or do not understand. This function is defeated if the lecturer does not take questions. Therefore, it is essential that some time be scheduled for questions and answers in the lecture itself, if the lecturer is not the quiz session instructor. In fact, even if the two are the same person, there is virtue in taking questions during the lecture. Because the questions are fresh, the lecturer gets immediate feedback and the whole science-learning process becomes more dynamic.

Teacher-Student Interactions

Of at least equal importance with what the science teacher does is how he or she does it. The way the teacher interacts with the students will have a profound effect on how they perceive their own ability to do science and, in fact, will affect their self-perceptions in other ways. (The American Association of Physics Teachers, recognizing this, has developed a training workshop for its members: "Instilling Student Confidence in Physics".)

Now let me switch to the pronoun "he," because what I will be addressing is the behavior of male science teachers. The fact that males predominate in science teaching is part of the science anxiety problem. I am personally most familiar with male behavior in science teaching since all my teachers were, and most of my colleagues are, men. It may be that female science teachers approach their profession differently, because of different expectations of girls than of boys. If they do not, then what I say in the next few paragraphs applies to them as well.

The fundamental issue, as we said earlier, is this: Does the science teacher see himself as a role model for students or as a member of an elite club? On this self-perception of the teacher hinges the level of anxiety in the science classroom. If the teacher gives the underlying message to the students: "Here is science; it is something I have learned to do. You can learn it too, because you are not very different from me," then the students will respond positively, with little or no anxiety about learning science. If on the other hand, the science teacher gives the message: "I am smarter than you; I was born smarter than you, and there is nothing you can do to comprehend science as well as I do," then the student will respond in kind. This sort of instructor is a source of science anxiety, and his approach to teaching is the scientist's version of machismo.

Science teachers communicate their attitudes in very concrete ways. The teacher who responds respectfully to all serious student questions, regardless of how simple they may seem, provides a good role model. The one who responds with sarcasm or by saying "We have no time for questions" or "I can't explain it any more simply" is an anxiety producer.

Now we must ask the same question about the teacher who causes science anxiety that we asked about the students who experience science anxiety: What is the root of this behavior? What we find is that there is one root that both the anxiety producer and the anxiety sufferer share in common: society's view of the scientist. Scientists grow up subject to the same influences as nonscientists; their interests and talents just make them respond somewhat differently. However, they can be victimized by the same stereotypes as the science anxious. Just as one young person viewing the media image of the scientist as an eccentric, poorly attired, unfeeling, computing machine in human form may respond by saying, "I can't or won't be like that," another may say, "I can and will be like that." We scientists, in other words, may buy into the image that has been constructed of us. Now this entails more than just dressing in beige corduroy jackets and Hush Puppies. It also involves establishing a pattern of how to get what some psychologists call "strokes"; i.e., how to feel rewarded and worthwhile. Unfortunately, the way that many of us do this is to internalize one of the negative self-statements so common among the science anxious: "My self worth is determined by how smart I am." If all our energies go into showing how smart we are in science, then it becomes threatening to believe that others may be as smart or as talented as we are. The scientist who is trapped in this psychological bind is the one who becomes the science anxiety producer; the one who agrees with antiscience humanities scholars that there is an artistic and a scientific mind, and that people have one or the other; the one who is uninterested in any students who are not majoring in his field; and the one who contributes to the further isolation of science from society. Harsh words, indeed, but they need to be said. We cannot afford the luxury of a scientific elite distant from society -- not while politicians and corporations are making technological decisions that affect the food we eat and the air we breathe.

It is important that the science-anxious reader understand the nature of the pressure to which the science teacher is subject. First, this understanding humanizes the teacher in the eyes of the student and demystifies the science teaching process (even the process of poor science teaching). Second, and even more important, it allows the science-anxious student to see the cause of "elitist" science teaching. It may very well be that a particular science teacher is unwilling or unable to be a role model. His habit of sarcastic answers or dismissal of questions may be too ingrained -- but the student need not take poor response on the part of the teacher as an indication of the student's lack of worth. Some teachers will always be the Typhoid Marys of science anxiety. I hope that these few paragraphs will provide a kind of innoculation against them.

What about the teacher who is genuinely trying to reach the students, to provide an open, inquiring atmosphere in the classroom, and to act as a role model for students? Even this sort of teacher must be careful to avoid certain pitfalls that can trigger anxiety responses in students. For example, we once had two women enrolled in our Clinic who were in the same science class with a very good and compassionate teacher. However, before every exam, in an effort to encourage them, he would say, "Now, I know you can get an A on this test." Well, his words had just the opposite effect from that which he intended. The students responded in the same way as to pressure from parents: They felt they couldn't live up to his expectations. Science teachers must tread carefully the road paved with good intentions -- we know what road that is.

Just as students are often unable to assess the quality of a science textbook, they are also unable to decide when an exam is good or bad. That is, they know if it was hard or easy, but not whether it actually tested what it was supposed to test. The science teacher, however, usually has a pretty good idea about the virtues and faults of a certain exam. If he thinks that a particular question was not a good one, and didn't test what it was supposed to, it is best that he be honest with the class about this. Otherwise they will spend their time trying to "psych him out" on the next exam, rather than trying to learn the material. The more balanced and grounded the students feel in the science class, the more they will be able to learn science.

Conversely, if the instructor thinks he has given a good question and the students have not learned the material, he should say so. In this way, they get a clear message about what they need to learn.

Sexism in Science Teaching

One compelling reason for using the pronoun "he" in describing science teachers in the preceding sections is the dearth of female scientists and science teachers. Although their numbers have increased somewhat in the last decade, due in large measure to the feminist movement and to women's enhanced perceptions of the possibilities for a satisfying life independent of or combined with marriage and family, the number of women in science teaching is still small. In some areas it has decreased as a percentage of the total. An article in *Physics Today* by Vera Kistiakowsky,[2] a physics professor at M.I.T., documents the rather small numbers of women in the sciences and the very slow rate of increase even with affirmative action. Women scientists and science teachers are hired less frequently, work at less prestigious institutions, and are paid less than their male counterparts for the same jobs. As we mentioned in earlier chapters, there is no substantive data indicating that women are in any way less endowed with the skills that make a good scientist. Rather, a variety of socialization processes operate to steer women away from the study of science. These include:

1. Pressure from parents and friends to conform to "traditional" women's roles: either marriage and family, or "women's careers," such as nursing, teaching, clerical work

2. Attitudes of teachers toward women; in particular, the tracking of women away from science as early as the third grade

3. Overt sexism in science and math textbooks: always showing men doing science and math, or even worse, portraying women as illogical, comical figures, unable to reason effectively

4. Sexist attitudes of male scientists toward their female counterparts. The woman scientist can expect to get little encouragement from her male coworkers. Since much of the reward in science

consists of the esteem of peers, women will tend to avoid a profession that is unrewarding in this sense.

There have been many studies in the past two decades documenting the ways in which society pressures women to avoid science.[3] Underlying this pressure is an attempt, not unlike other attempts, to keep a group of people powerless. To be able to do science is to have power, to have a sense that you can master the secrets of the universe, that you can understand technical arguments. Denying women this power is like denying them the vote. In fact, on political issues with a scientific base, such as pollution and nuclear safety, it is denying them the vote, because it denies them access to the tools needed to make an informed decision.

The vicious cycle of few female scientists as role models leading to young girls' low opinion of their abilities to do science, leading to few female scientists, continues. However, a study by Marcia Guttentag and Helen Bray [4] of girls in grades ranging from kindergarten to high school showed they were willing and able to change their self-perceptions and see themselves as capable of doing science, if their science teachers were committed to teaching science in a nonsexist way. The role of the science teacher was crucial in changing attitudes.

Let us, at this point, spend some time considering how to remove sexism from science teaching. The first place to look is in the science textbook. Many texts show only boys or men doing experiments. This negative message is not lost on the girls in the class. A few newer textbooks, however, have begun to show women and minorities in their examples of students doing science. The science teacher should seek out these books and attempt to use them in the course. Obviously, this should not be the only consideration in choosing a text. A poorly written but nonsexist text is of no use to anyone. However, there are now enough new texts on the market that there is a fair choice of good, nonsexist textbooks.

By the time they have reached high school, many girls have begun to feel so uncomfortable about science that they hesitate to ask questions in class. Once this pattern is established, it is hard to break, unless the teacher explicitly confronts the issue. The usual interpretation for girls' keeping a low profile in science is that they don't want to appear smarter than boys because it will hurt their social interactions. This is true in some cases, but in my experience with science-anxious college women, the majority of them have come to believe that they are not as smart as men and that any questions they would ask in class would be stupid ones. What I do in the classes I teach, therefore, is to raise the issue in the first week, just as the pattern of who asks questions is being established. I point out that the women are holding back, that this is a result of sexism, that if they are interested in having a fruitful and pleasurable science-learning experience, then they must start asking questions. I reassure them that all questions will be treated with respect and seriousness and that only by questioning will they learn. I usually quote to them something a professor of mine, a very well known physicist, said when I took a course from him in graduate school: "There are no dumb questions. There are only dumb answers!" And of course the teacher is in charge of those.

It should go without saying, but let me say it anyway: Overtly sexist comments, jokes, transparencies, etc. do untold damage to women who are studying science and must be avoided.

In the science laboratory, the instructor must be vigilant in recognizing problems of gender stereotyping when they occur. If two lab partners are of opposite gender, they must share equally in the tasks at hand. The old pattern of boy-works-with-equipment and girl-writes-down-numbers must be broken. Some of the women in a science lab will need extra attention, because they will have had less hands-on experience with tools in general. They must get this extra help.

In the last decade we have seen an extremely positive phenomenon in colleges: the increased enrollment of older women who had deferred their career and educational aspirations while raising a family. A good number of these women take courses, or hope to take courses, in science. They are mature, serious, hard-working students who are often attempting to overcome their own antiscience backgrounds. They must be strongly supported in this endeavor by their science teachers. If we help them overcome science avoidance and science anxiety, then we affect not only them but their daughters -- and there is no better two-for-the-price-of-one bargain that I know of.

There are other, subtler issues involving women and science that we must mention. There has been some discussion of late among social scientists as to whether the "scientific paradigm," that is, the scientific way of looking at things, is intrinsically male. Perhaps, the argument goes, the very questions scientists ask and the way they do research to find the answers fit the traditional male role model: lack of emotion, interest in "things" rather than in "people," aggressive competition in research as in sports, and so on. Other social scientists have pointed out, on the other hand, that the perception of the scientist is in some ways the antithesis of the male stereotype: the scientist is thoughtful rather than physical, concerned with detail, intuitive -- in a word, conforming to some of the traditional female roles. I am personally not sure where this discussion can lead or what if any answers there are to these questions. Perhaps all other things being equal, a greater number of men than women will freely choose to become scientists or to study science, just as different cultures, for example, place different emphases on art and science. It is the task of science teachers to strive to make sure that all other things are equal.

What Parents Can Do

The strongest influences in a young person's life are his parents and teachers. Trite but true. Parents can therefore play a strong positive role in their child's science learning experience. There are a number of ways they can help their child. In the first place, the parent, whether or not he or she has had a good experience (or any experience) with science, is much more aware than is the child of the importance of science in our lives. Parents can encourage the child when the going gets a little tough. Parents can also monitor the child's science learning just as they monitor the child's learning in other areas. You do not need to know science yourself to elicit from your child whether he or she is excited about the science class, whether he or she thinks that the teacher is effective and compassionate, whether girls and boys are treated equally. Look at the textbooks. Are they up-to-date or 10 years old? Things move fast in science -- an old textbook is useless. One of my nephews when he was in the fifth grade was rewarded by his teacher for his good performance in science. She gave him a book on magnetism written 15 years earlier. Her intentions were good, but what he wound up with was a compendium of misinformation and outdated concepts.

The parent should also visit the science laboratories in the child's school. This is especially useful in the high school. Ask that a tour of the labs be part of the yearly open-school-week parent-teacher conferences. Find out how old the equipment is. See if the lab book of experiments is up-to-date. Ask how many students share one lab setup. (The answer should be no more than three.)

But parents must do more than monitor the child's science classes. They must monitor their own parental expectations. These expectations may on the surface have little to do with science, but they will almost certainly affect the child's level of anxiety about science. In our experience with science-anxious students, we discovered a remarkable thing: Science anxiety was often an indicator of other problems the student had. Typical was the male student whose parents wanted him to be a doctor. Unwilling or unable to confront them with his lack of desire to fulfill their expectations, he found a way out: science anxiety. One sure way not to get into medical school is to fail a science course or two. I am by no means suggesting that the student did this consciously. Rather, the psychological pressure found a socially acceptable outlet: poor science performance. Failing, for example, English composition does not have nearly the desired effect. But failing chemistry, which everyone believes is too hard anyway, is a way out of the situation for this student.

Parents with the best of intentions can often give negative messages to their children. "We know you can do it, if you only put your mind to it"; "The whole family expects great things of you"; "You can do anything you set your mind to." Statements such as these only increase the pressure on the student. Parents, if they really want to help their children, must look for the kinds of coping statements that we talked about in the previous chapters: realistic, objective descriptions of their child's situation and prescriptions about how he or she can feel more comfortable. "Where do you think you are having trouble? Why don't you find other students in the same class with whom you can work?" These kinds of questions show real concern for the child's situation and can help him or her overcome science anxiety.

The final thing the parents can do is to combat societal prejudices against science. There are many negative influences to counteract, as we have seen: "Mad Scientist" movies, prejudice against women, media stereotypes of scientists, and pseudoscientific quacks, to mention a few. If you have read the preceding chapters of this book, you have a pretty good idea of what to be on the lookout for as a parent. In addition, the parents can introduce a child to science in several ways: By giving chemistry, biology, or physics kits as gifts, by encouraging children to watch science shows on TV, such as "NOVA" or "Contact," by frequent trips to museums, and so on. Even the parent who has had bad experiences with science will benefit. Through the child, the parent may recapture the joyful curiosity that has been lost to societal prejudice, bad teaching, and science anxiety.

[1]C.F. Gauld, "Physics Teaching and Cognitive Functioning -- A Neo-Piagetian Perspective," *The Physics Teacher*, 513-518, November 1979.

[2]V. Kistiakowsky, "Women in Physics: Unnecessary, Injurious, and Out of Place?" *Physics Today* 33:32-40, February 1980.

[3]See the Bibliography of Women and Science at the end of the book.

[4]M. Guttentag and H. Bray, *Undoing Sex Stereotypes: Research and Resources for Educators*, New York: McGraw-Hill, 1976.

6

The Science Anxiety Clinic

In this chapter we shall consider how to overcome fear of science by "going inside" a Science Anxiety Clinic; that is, by seeing exactly what is done to help students fight their fears about science courses and about science in general.

How the Science Anxiety Clinic Works

Our Science Anxiety Clinic at Loyola University of Chicago was the first such clinic in existence, dating from early 1977. As of this writing, a number of other science anxiety programs have sprung up across the country. I shall of course describe what we do in our Clinic; the procedures vary from school to school depending on the nature of the student body. What we have observed from the students who have taken our Clinic program (nearly 200), what they have reported to us, and what we have been able to measure constitute the foundation of the conclusions that are drawn in this chapter. I hope that psychologists who read this chapter will find our observations a good starting point for detailed research into the phenomenon of science anxiety.

Who are the students who come to the Clinic? How do they find out about its activities, and how do we select them? At the beginning of each school semester, we post fliers announcing the Clinic. A typical flier is shown in Figure 8. In addition, we place ads in the Loyola student newspaper. This, we hope, will reach the whole campus and attract students from various disciplines, since every student at Loyola must take at least one year of a science.

Do You Have
Science Anxiety?

*Are you avoiding taking science because of prior bad experiences, or because you think it's beyond you?

*Are you limiting your career choices by not taking science?

*Are you taking science, but feel anxious about it?

Join Loyola's Science Anxiety Clinic

Figure 8. A typical flier used to advertise our Science Anxiety Clinic.

Once the announcements have been out for awhile, we go into certain classes personally to announce the availability of the Science Anxiety Clinic and to answer any questions students may have. These classes include introductory classes in the sciences and the social sciences. Although we expect that students in the liberal arts and students who are not already enrolled in a science course will come to the Clinic, we realize that those students who are already taking science are "up against it" and have the most at stake. Since the students in science major programs or premed programs or such social sciences as psychology will need strong science preparation, we approach them directly in their classes. In addition, we send an announcement to all department chairpersons asking them to have their instructors announce the Science Anxiety Clinic in their lectures.

This procedure may look a little contradictory. After all, if we are trying to attract students to science, wouldn't it be better to recruit more strongly among students who are not taking science? Why recruit among students in science courses? Aren't they free of the symptoms of science anxiety?

In fact, what we discovered very early in the life of the Clinic was that many students even in the sciences are very science anxious. Professors of science generally recognize this fact when we point it out to them. They have wondered about the same thing I wondered about: Why do bright students fail in science? It is precisely the science classes where we are most welcome, since it is in the interests of the science teachers to have their students succeed. Furthermore, these students, who see a very close relation between their success in science and their long-range goals, such as medicine or dentistry, are most eager to enroll in the Science Anxiety Clinic so that they may conquer their fear of science as early as possible in their college career.

This is not to say that we don't get liberal arts students in the Clinic; in fact, we do. They are not, however, in the majority. (One of our goals for the future is to run Clinic groups specifically geared to the needs of liberal arts majors.)

I was initially surprised at the number of science students who signed up for the Clinic. If they were already studying science, why were they anxious?

I shouldn't have been so surprised. I had a good dose of science anxiety myself as a college student. In my case, I was terrified that I couldn't grasp physics. This was a rather painful anxiety, since physics was my major. I actually loved the subject, but I froze every time I had to take an exam in it. One of the motivations for starting the Science Anxiety Clinic was my own resolve that no one should have to go through what I went through as an undergraduate. And the large number of students who, although majoring in science, enroll in the Clinic have shown me that not much has changed since I was in their shoes.

Actually, the students are not usually anxious about the science that is their major, although sometimes this is the case. More frequently, the biology majors are terrified of chemistry or physics, and conversely. Since each of these majors requires courses in other sciences, the students could not simply turn their anxiety into avoidance. Thus they came to us.

Of the students who apply for admission to the Science Anxiety Clinic, women outnumber men by about two to one. This comes about, we think, for two reasons. The first is that women are taught to believe that they cannot do science as well as men. Each difficulty they encounter in a science course is therefore magnified for them and substantiates their negative self-image. Furthermore, they are generally the minority in science courses. The ratio of women to men is about one to one in introductory biology classes, somewhat less (about 40 percent) in introductory chemistry classes, and a good deal lower in basic physics courses, reflecting traditional stereotypes about the difficulty of the subjects and, therefore, the ability of women to comprehend them. As the level of the courses in all three of the sciences progresses, the ratio of women goes systematically down. Women are the primary casualties in the drop-out rate from science. So we get a good number of women in the Clinic who are taking introductory science and already feeling anxious about their ability to succeed.

The second reason women outnumber men as applicants to the Science Anxiety Clinic is that men are socialized to believe they are naturally better at science; thus, they are ashamed to ask for help when they run into difficulty. We suspect that fewer science-anxious men are actually willing to admit their anxiety than are their female counterparts. Hence the disparate numbers of men and women in the Clinic.

Once students have applied to join the Clinic, we do some preliminary screening. Since the tight schedules of the Clinic leaders only allow for sessions at rather specific times during the week, we are forced to eliminate some students simply on the basis of schedule. We encourage them to reapply the following semester, at which time they get first priority. We also attempt to meet their needs with some informal counseling and perhaps a mini-Clinic session.

The remaining students are given a brief questionnaire to fill out. A copy of it appears at the end of Chapter 3. We attempt to ascertain from this questionnaire whether the students are specifically anxious about science or whether they have other problems which they need to deal with in other ways.

An interesting aspect of science anxiety seems to be its ability to act in some cases as a kind of early warning system for other problems. More will be said about this later. We mention it now since some of the students who apply to our clinic are actually suffering from personal problems that manifest themselves as anxiety about science. A student with problems at home may wind up flunking chemistry and come to our Clinic. We try to screen for this, that is, to detect the cry for help underlying the application to the Science Anxiety Clinic. We encourage these students to avail themselves of the other types of counseling available at the Counseling Center.

We run the Clinic for seven weeks, with each weekly session lasting an hour and a half. At the end of that time we interview the students about their experience in the Clinic and consider their various comments when we plan for the next semester.

At Loyola, students who are accepted into the Clinic begin their sessions a few weeks into the semester and finish just before final exams. A typical Clinic group contains six to 10 students. There are two group leaders. One is a professor or graduate student from one of the science departments. The other is a psychologist from the staff of the Counseling Center or a graduate student in psychology or counseling. In these group sessions students do three things: They learn the skills needed to study science; they explore the roots of their science anxiety and devise ways to cope with it; and they learn some techniques for relaxation that they then use to "desensitize" themselves to science-related situations that may produce anxiety.

Each of the seven sessions is a mixture of these various parts. Under "Science Skills" we deal with how to take notes in a science lecture;

how to work problems in chemistry, physics, or astronomy; how to deal with graphs and equations; how to function effectively in a laboratory; and how to read a science textbook or a popular article about science. Under "Exploring and Coping With Science Anxiety" we include discovering the things we tell ourselves that keep us from learning science and what new things we can say to replace the old, bad self-messages. The third component, "Relaxation-Desensitization," includes learning how to tense and relax various muscle groups in the body on command and getting in touch with what physical signs we give ourselves to show we are anxious (cold sweat, upset stomach, foot-tapping, and so forth). It also includes the development of a "hierarchy" of science-anxiety-producing situations, ranging from looking at the syllabus in a science course to working problems on the final exam. We learn how to deal with these situations in a relaxed way; that is, we desensitize ourselves to the anxiety producing potential of science.

Different members of the group will respond differently to the various components of the Clinic. Some need the skills training most; others respond to the desensitization or to the exploration of self-messages. We have observed that the combination of all three is an effective mixture for reducing science anxiety in the vast majority of students who come to our Clinic.

At this point, the reader who is perhaps knowledgeable about the operation of math anxiety clinics will notice that this all sounds familiar. Indeed, the Science Anxiety Clinic follows the model of the math anxiety clinics.

Let us now turn to a detailed description of what actually happens in the Science Anxiety Clinic. As you read these sections, look for situations that seem familiar to you, particularly if you number yourself among the science anxious. Science anxiety is a fear shared by many; as such, it contains elements common to many. It is very likely that you will see some of your own anxieties alluded to in the following description of our Clinic procedures.

Session 1

At the beginning of the first session, the group leaders, a scientist and a psychologist/counselor, introduce themselves to the group and have the group members, some six to 10 students, briefly introduce themselves to one another. The scientist discusses how he or she became involved in the Clinic. This discussion can include a brief description of the scientist's own experiences with science; in particular, any unpleasant or anxiety producing experiences. This is not as bizarre as it may sound. Scientists have not had uniformly good experiences with science. In fact, they have had the same school experiences as anyone else and have gone on to become scientists because of some of these experiences in spite of others. Even if their experiences in learning their own science have been uniformly good (which is highly unlikely), they can probably relate to some of the students' experiences in the other sciences. A physicist, for example, may not have had a very pleasant time with chemistry. If this is the case, it is important that this be communicated to the students, since one of the most important functions that the scientist plays in the Clinic group is that of role model. Therefore, any experiences that parallel those of the students should be emphasized. In my own case, I was a victim of science anxiety in physics, which was my chosen major, and which is now my profession. Revealing this to the students in the Clinic makes a very strong impression and immediately communicates to them that they can overcome their fears about science. This is particularly important to those who are majoring in science and who therefore have a great deal at stake in overcoming their anxiety.

The psychologist/counselor then makes a similar introduction, indicating the reasons for his or her involvement in the Clinic and discussing personal experiences with science. Some of the psychologists with whom I have worked have had more than a brush with science anxiety. They are interested not only in helping the students but also in exploring the roots of science anxiety in their own cases. This does not by any means detract from their value to the Clinic; on the contrary, it adds to their motivation and draws them closer to the students. Anyone who has suffered from science anxiety can empathize with those who are suffering from it now.

One important condition for a successful Clinic is that the group leaders consist of one of each gender. Since science anxiety is in large part a result of social conditioning, it counts more women than men among its victims. It is therefore important that the female students in the Clinic groups see at least one female role model, either the scientist or the psychologist. I do not think a group led by two males can be nearly as helpful to its female members. Since there are more male than female scientists, the usual, but by no means ideal, situation is that the scientist is male and the psychologist is female. We have been able to run a number of groups with female scientists and male psychologists. The groups appear to have been very successful; the reversal of traditional role models left strong positive impressions on the students. They have now seen that it is OK for women to opt to work with "things" and for men to opt to work with people. This gives them permission to reassess their own career goals.

After the group leaders have given their own brief "science autobiographies," and before the Clinic students do the same, the group leaders outline the rules of the group. These rules should also be handed out in writing to the group members. The rules include the following:

1. The students are given "permission and protection." That is, the group members make a verbal contract with one another that what goes on in the group remains confidential. This protection then gives the group members permission to reveal themselves to one another and to frankly discuss their problems with science.

2. The group members are asked to use so-called "I-statements" when discussing their science anxiety; that is, they are asked to always speak in the first person. This rather simple rule is extremely important because anxious students will try to put some distance between themselves and their fears by depersonalizing their descriptions of them. For example, a student describing his or her feelings when confronted with a difficult homework problem in chemistry might say something like: "When you look at one of these problems, and you see that it's long and looks complicated, you begin to think that you'll never be able to solve it in a million years, and then you begin to think that maybe you shouldn't be studying science..." The group leaders must have the student completely restate the issue as follows: "When I look at one of these problems, and see that it looks long and complicated, I begin to think that I'll never be able to solve it in a million years, and I then begin to think that maybe I shouldn't be studying science." This apparently simple restatement is crucial to the student's facing this as his or her own problem. As people who have worked with groups know, it is no simple task to get the group members to personalize and face their problems head-on. It is, however, essential that they do so. So the "I-statement" rule is one that we insist upon in the Science Anxiety Clinic.

3. The other rules have to do with commitment to the group and to the seriousness of the endeavor. The group leaders must insist on punctuality and on no unexcused absences. It becomes very tempting for students to skip the group from time to time. Sometimes the group may be meeting the day before a midterm exam. Sometimes a student may have confronted one of his or her problems with science anxiety in a group session and may be frightened and resistant to coming to the next group. For the sake of the students, and the success of the group as a whole, these absences must not be allowed to happen. Group cohesion is an essential part of the process. One of the strongest weapons that a student has to combat science anxiety is the knowledge that there are others in the same boat. The group provides a continual reminder that this is the case, but repeated absences sabotage the process.

After the Clinic rules have been explained, the students are asked to present their own science autobiographies. It is our practice to tape-record the first few sessions, so that the staff psychologist who oversees the Clinic can refer to the tapes during discussions by group leaders of the weekly sessions. We ask the students for their permission to tape, with the proviso that if at any time a student does not wish to be recorded, the group leader will turn the tape machine off.

The students present their science autobiographies. They begin by telling the group what specifically brought them to the Science Anxiety Clinic. They then go on to discuss their earlier experiences with science and how they think these may have affected their attitudes toward the study of science. While each student is reporting, the other students are encouraged to speak up if they recognize common experiences with the speaker. They find that all of them, although unique, share a wide range of common experience about studying science and fearing it. Some of the questions that the students address are:

> What were my experiences with science in grade school? How did I feel about them?

> What were my experiences with science in high school? How did I feel about them?

> What experiences am I now having in college with science? How do I feel about them? How is science in college the same as or different from before?

> What are my family's attitudes toward science? Toward people who study science?

> What are the attitudes of my friends and people I grew up with when I tell them that I am thinking of taking a science course? Of majoring in a science?

> Were the attitudes of my family and friends different for males than for females?

> When do I feel most fearful about science? During a lecture? During an exam? While doing homework? When talking to another student in the class? When considering asking the professor a question? When considering registering for a science course?

> What image comes to mind when I hear the word "scientist?"

> How do I think the study of science might affect my career goals?

> How do I think the study of science might affect my attitude toward myself?

These questions help spark the students' thinking about their science anxiety. Of course, these questions are not meant to limit the discussion. Nor do we attempt to answer them at this time. The answers are discovered by the students themselves in the course of the seven-week Clinic program. As the students talk, the group leaders help move the discussion along, pointing out things that the various group members may have in common that they have not already realized. The discussion usually takes 30 to 45 minutes. Each student must present his or her science autobiography before the group can move on.

Here are some typical statements we might hear from students during their presentation:

> In high school I had a chemistry course where all the teacher did was dazzle us with fancy demonstrations while not actually teaching us anything.

> I want to be a doctor, but since I'm a girl, my family thinks I should be a nurse.

> I'm a literature major. That's where my talents lie. I think I'd like to know some science, and I think I should know it, but I'm afraid it's beyond me.

> I always had trouble with word problems in math, and now I have them in math and science.

> I'm afraid to ask a question in science class because everyone else will think I'm dumb.

> Only superbrains can understand physics.

> Science in college is completely different from science in high school.

> I'm a psychology major. I know I have to take a biology course but I don't think I can pass it. And I wouldn't even attempt chemistry or physics.

> I don't know what to think about nuclear reactors. I wish I could understand a little about how they work.

> I'm afraid to ask a question in biology because I think the professor will put me down.

> I try to read my science book, but I don't understand anything!

> I want to be a nurse, but I have to take chemistry, and I'm terrified!

These statements and statements like them make up the bulk of the science autobiographies during the first Clinic session. A typical presentation might go as follows:

Student A, a male, says: "Every time I walk into chemistry class, I feel sick. I'd like to go to medical school. So I know I have to take chemistry, but I really hate it. I just know I can't do well. I think that everyone in the class understands what's going on, except me."

125

Student B interrupts, "That's exactly the way I feel in my physics class!"

The psychologist asks Student A, "When did you first start to feel this way? Does this go back to high school?"

"No," replies student A, "I took chemistry in high school, and it was trivial. All the teacher did was mix chemicals and make little explosions in the lab."

"Do you think you learned any chemistry in high school?" asks the scientist.

"No," replies student A, "And that's part of the reason why I'm having trouble now, I think everyone in class had a decent chem course in high school, and they all know more than me. So there's no way I can catch up no matter how hard I study."

At this point, the student has begun to reveal some of the hidden messages that underlie his science anxiety: the fear that he has not been taught the skills to do science (a fear that is based in reality) and the fear that he will never be able to acquire those skills (an unrealistic fear). The Clinic will deal with both these fears: the first by providing him with science skills and the second by dealing with the unreal nature of statements that say someone can never attain a reasonable goal.

The third bit of information that the group leaders elicit from student A is how he knows when he is anxious:

"What tips you off that you are feeling anxiety?" asks the psychologist.

Student A replies, "My hands begin to sweat."

Student B chimes in, "I get pains in my stomach."

Student C says, "In my case, I suddenly notice that I'm tapping my foot and I can't concentrate on my studies."

Once having obtained information from the students about their science skills, their messages to themselves about their ability to do science, and their ways of telling when they are feeling anxious, the group is ready to begin working to reduce the science anxiety of its members.

Before moving on to a discussion of how this is accomplished, we should note a number of things about the science autobiography stage of the Clinic. First of all, things don't necessarily go as smoothly as outlined in the example just presented. Students have been suffering with this fear for awhile, and it is not easy to begin to face up to it, especially in a group of people they have just met. It is extremely important for the group leaders to make the Clinic a safe and comfortable place from the outset. This is the main reason why it is good for the scientist and psychologist to reveal their own problems with science. It then makes it easier for the students to follow suit.

Second, the fact that all the students are pretty much in the same boat is a giant first step toward reducing science anxiety. Each student who comes to the Clinic believes he or she is the only one truly afflicted and that the anxiety experience is unique. Just hearing another student say, "I feel the same way" is a revelation that takes away some of the fear of having to fight this thing alone.

A third observation is that not all the students know when they are feeling anxious. They have in many cases blocked out these feelings. It is essential the students be helped to recognize when they are feeling anxious, otherwise it sneaks up on them and distracts them from functioning well in science virtually without their realizing what is happening. And it is not enough to find out that, say, they feel scared during an exam. Science anxiety is not simply test anxiety. The fear is there long before the exam, and science-anxious students must learn to recognize their own symptoms of anxiety that they experience in their daily contacts with science. These may be quite subtle and appear to have little to do with science. Loss of appetite, insomnia, and nail-biting are all examples of anxiety related behavior. If the students exhibit any of these or similar behaviors, the group must try to see if it is related to their experiences with science. In order to reduce their anxiety, they have to recognize when they are feeling anxious.

After the students have presented their science autobiographies, the group leaders outline the procedures that will be used to reduce science anxiety. These fall into the three categories mentioned previously:

1. Science skills
2. Cognitive restructuring
3. Relaxation and desensitization

Each Clinic session is made up of the three parts in about equal proportion. Clinic sessions are one and a half hours. Therefore, each of the three parts is given about half an hour. This is, however, not a rigid rule; we try to "go with the flow." If the students are making good progress in dealing, say, with their negative messages to themselves about their ability to do science, we don't cut them off to work a science problem on the blackboard.

The first of the categories, Science Skills, includes teaching the students how to learn science. The fact that science is learned differently from other subjects is almost never addressed by science teachers. Students are somehow expected to "pick up" the necessary skills. Some do. Some do not and this leads to anxiety. Some have had such bad experiences with learning science skills in high school that they are trying to avoid science completely in college. Since this is not possible, at least at our school, they come to the Clinic to learn some skills and to reduce their anxieties.

The scientist tells the group that science skills include:

1. How to read a science book and how this is different from reading a nonscience book.

2. How to listen to a science lecture and how to take notes without losing the train of thought.
3. How to go about setting up and solving science word problems.
4. How to comprehend equations ("formulas") and what role they play in science.
5. How to comprehend graphs and what role they play in science.
6. How to study for exams and how to take exams efficiently in a science course.
7. How to function in a science laboratory to get the most out of it.

In each of our Clinic sessions, we try to deal with one of these skills and to practice it in later sessions. (Naturally, if some are not a problem, we don't bother with them. We have had groups where no one has had any anxiety about science labs. In that case, why waste time dealing with it?)

The second category of tasks we undertake in the Clinic goes by the name Cognitive Restructuring. It simply means that the students learn to explore the cognitive, or thinking, levels of their anxiety, the things they tell themselves, and to replace the negative, scary messages with more positive, objective ones. The psychologist explains: "As I've heard you talking, you've reported a variety of individual experiences in which you feel anxious and react emotionally. For instance, it seems that it becomes difficult for you to give full attention to what you are doing. You become focused on yourself; that is, your thoughts shift to an internal emphasis and it becomes difficult to focus on the task, such as homework problems or exams.

"Much of what you say to yourself is *self-evaluative and self-critical*. These are what we call *negative self-statements or self-thoughts*. They seem to get in the way of what you have to do. Self-statements often include some expectation that some horrible consequences (catastrophe or negative outcome) will occur. The more we dwell on these consequences, the less attention we can focus on our task, whether it be studying, performing an experiment, or whatever.

"We can learn to *change* negative self-statements to positive ones, so that thinking doesn't get in the way of what we want to do. *Positive self-statements* allow us to focus attention outside ourselves, that is, on the task at hand. We can relax with ourselves and be more attentive to other things.

"The first step toward controlling your thoughts and where you focus your attention is becoming aware of *when* you are producing negative self-statements. When each of us can recognize these statements in ourselves, it is the first step in changing.

"Whenever you find yourself getting upset, ask yourself what you are saying to yourself. Often the statements take forms such as 'Isn't it terrible...' or 'Wouldn't it be awful if...?' For example, 'Isn't it terrible that everyone knows more physics than I do' or 'Wouldn't it be awful if I fail this biology exam, because if I then fail the next one, I'll never get to medical school?'

"The recognition of such thoughts will be the *cue* or *bell-ringer* for us to produce different thought that will counteract the negative self-statements and allow attention to shift to the task at hand. We will learn the specific ways to do this over the next six sessions -- through practice in the group as well as at home. We will focus on challenging negative self-statements."

The psychologist then explains the third component of the Clinic: Relaxation and Desensitization:

"In addition, you have talked about physical tension associated with studying or work regarding science courses. This tension can take many forms such as tightening in the stomach or neck, sweaty palms, pounding hearts, heavy breathing.

"We are going to work directly on controlling your body tension by learning how to *systematically* relax all portions of your body. The purpose of *relaxation training* is that the muscles in your body can't be relaxed and tensed at the same time. Thus once you have learned relaxation, you can counter tension and anxiety as you experience it in your science courses and elsewhere.

"The specific technique we will be using in conjunction with relaxation training is called *desensitization*. Once you have learned the relaxation technique, desensitization can be used to counter anxiety, tension, and feelings such as you experience in science-related situations.

"The way in which we will do this is to determine specifically the science situations in which you become progressively more anxious, building a hierarchy from the least to the most anxious situations. Then as you learn the techniques of progressive relaxation we will start the desensitization. This is done by having you repeatedly *imagine* the specific situations from the science anxiety hierarchy while you are relaxed. By having you imagine these very briefly, while you are deeply relaxed, the science situations that normally arouse anxiety gradually become less potent, so that they no longer make you anxious. We start with those science situations that bother you the least and gradually work up to those that bother you the most. You can't be anxious and relaxed at the same time, so essentially we are substituting your *learned anxiety responses* with *learned relaxation responses*. We are replacing old responses with new ones.

"The anxiety hierarchy is a major part of the desensitization technique. Our objective is to determine the situations related to science or science courses that cause you slight, controllable amounts of anxiety all the way to strong panic responses. It is not necessary to determine every instance, since generalization from one instance to another will bridge the gap.

"Here is a model of a hierarchy for test anxiety. For next week's session list 10 to 15 of your personal situations related to science or a science course in which you experience anxiety. Remember to order situations from least to most anxiety. Be specific regarding place (where you are), time (time of day or night), people (other persons in the situation),

---

activity (what exactly you are doing -- that is, studying, sitting in class, and so forth."

Sample Science Anxiety Hierarchy

1. Deciding which instructor to take for a course.
2. Sitting in class on the first day.
3. Not being able to understand the homework.
4. Thinking about how you have fallen behind in reading.
5. Sitting in lecture and knowing you do not understand the material.
6. Going after class to the instructor with a question you think you should know the answer to.
7. Asking a question in class when no one else does.
8. Giving an oral presentation in class.
9. Cramming the night before an exam.
10. Studying immediately before an exam.
11. Discussing an exam with classmates right before taking it and feeling they know more than you do.
12. Not being able to work the first problem on an exam.
13. Realizing halfway through the exam that you do not have enough time.
14. Being distracted by something during the test and not being able to concentrate at all.

"The reason we want you to be specific is that we shall construct a composite hierarchy from the individual hierarchies that you each present. Since you will be asked to visualize the various science-related scenes in your hierarchy as clearly as possible, you need to describe them in detail."

As you can see, the relaxation-desensitization procedure is a kind of conditioning. The students learn to distinguish clearly between tension and relaxation (which is not a clear distinction for many people, especially those prone to anxiety). Then they learn to relax various groups of muscles on command. Once they have mastered this, they are conditioned to relax their muscles in the presence of science anxiety producing situations: lectures, homework, exams, and so forth. This newly formed skill -- relaxation on command -- carries over into their science classes.

Those readers who are familiar with this systematic desensitization technique will recognize that we are doing something a little unusual here. Systematic desensitization has been a much used technique for various kinds of anxiety reduction since it was developed by Edmund Jacobson some decades ago[1]. However, it is usually used on an individual basis. The person who is anxious develops an anxiety hierarchy; the psychologist works with this person alone to desensitize him or her to the hierarchy of anxiety producing items. In our Clinic, we do group desensitization. We have seen that the students in our Clinic share many of the same anxieties (not surprisingly, since they are all fearful of science). So when they bring

Reprinted with permission from *Chemistry* (now *SciQuest*), pp. 6-9, October 1978.
Copyright 1978 by American Chemical Society.

in their individual hierarchies, we look for elements common to several of them. We then construct a composite close enough to the students' individual items that they can visualize the scenes with no difficulty and can therefore be desensitized. Of course this could only work in a group focused on a limited theme, such as science anxiety.

After the psychologist and the scientist have outlined what the Clinic sessions will consist of and what the group will accomplish in the seven weeks, homework is assigned. The students are asked to make up their own science anxiety hierarchies and bring them to the second session. Each hierarchy should consist of 10 to 15 items, ordered by the level of anxiety they provoke. The psychologist explains: "The best way to do this is to write down all the items on separate index cards and then to look them over and decide on an order. It may not be immediately obvious to you which items belong where until you think about it. Therefore the index card format allows you to shuffle the order until it feels right to you."

The second homework assignment is a list of science readings. The students are asked to read short excerpts from various sources and are told they will be quizzed on the content in the next session of the Clinic.

The final homework assignment is for the students to begin to note and record negative self-statements. They should also be asking themselves such questions as:

What negative thoughts am I having?

What triggered these negative thoughts?

Are these thoughts leading me to be anxious or blocked from functioning?.

When do these thoughts occur? All the time or in specific settings?

Where was I when these thoughts occurred?

Who was present?

What happened after I became aware of these thoughts?

What physical sensation did I experience that indicated I was having these thoughts? How do I feel anxiety physically?

After the homework is assigned, the leaders ask the group members if they have any questions or comments about the Clinic so far. These are dealt with briefly, and the first session ends.

The science readings the students are assigned for homework might typically include:

1. An article on biological effects of radiation.
2. An essay on black holes.
3. A typical page on optics from a physics textbook used for a freshman college course.
4. A page or two from a physics text used for a course for liberal arts students, containing a brief biography of Albert Einstein.

Session 2

The second session of the Clinic begins with a brief discussion of the homework assignments on listening to oneself. This is followed immediately by a quiz on the science readings:

1. What is the nature of the danger caused by radioactive strontium?
2. What is the danger of other radioactive isotopes?
3. At the point when a star becomes a black hole, what is the escape velocity of an object from that star?
4. What characteristics of a star determine whether it will become a white dwarf, a neutron star, or a black hole?

5. What is illuminance?
6. How does intensity of light vary with distance from the light source?
7. What did Einstein do for a living while he was publishing his epochal papers?
8. Name two areas of research to which Einstein made seminal contributions.

The students are given five minutes to answer questions about the readings. They are given no assistance by the group leaders. The scientist walks around during the quiz and looks at the students' papers as they are writing. In a word, we try to duplicate as closely as possible actual conditions during a classroom quiz.

At the end of five minutes, the scientist collects the papers and briefly looks over each one. The psychologist notes student responses during the quiz and after. When the scientist has finished looking over the papers, the group leaders initiate a discussion of the students' feelings about the quiz. It is essential at this time that the answers to the quiz *not* be given. The main purpose of this exercise is to trace for group members the incidence and timing of their anxieties. In particular, we hope that the students are able to focus on some of their negative self-statements during the quiz. The leaders ask such questions as:

What feelings were you aware of during the quiz?

What feelings are you having now that the quiz is over?

When were the feelings most intense?

Can you identify anything you might have told yourself during the quiz or after?

How were your thoughts about yourself related to your feelings?

Responses to these types of questions vary widely. For some students, the quiz is indeed an anxiety producing experience. They tell us, "I know this doesn't really count for anything, but as soon as I sat down to do the quiz I felt exactly as if I were in a science class." Others say, "I wish I'd done the readings more carefully and studied harder for the quiz." Some tell the scientist, "I really freaked out when you walked around and looked at my paper; I'm afraid of what you'll think of me." Still others volunteer, "I knew this didn't really count for a grade, so I don't care whether I did badly."

Each of these responses is a kind of code, which contains the kernel of the students' negative self-statements. The task of the group is to interpret the code and learn how to counter these statements with more objective messages. This "cognitive restructuring" will lead to reduction of science anxiety.

Notice, however, that even while this discussion is going on, there is still a serious source of tension in the group: The scientist has not yet presented the answers to the quiz. Although none of the students comes right out and asks, each is waiting with baited breath to see how he or she did. And, should the group leaders attempt to move on to other business without giving the answers, they will eventually be stopped by the students. What is going on in this interaction? The students have a good deal of their self-image wrapped up in how they perform on exams, since this has been the major source of reward since they started grammar school. Nevertheless, they don't want to appear too "grade conscious." So each one hopes that someone else will ask for the answers.

Before the scientist gives the answers, the students are asked to give themselves the grade they deserve on the quiz. After this has been done, the answers are presented. And now a second discussion ensues. How do the students feel now that they know how they did on the exam? Did they do as well as they thought? Did they do better or worse? What new things do they tell themselves once they know their exam scores? How do they feel about their grade relative to the other students in the Clinic?

What we have produced on a small scale is the environment in which these students operate daily. We assist them in focusing on the feelings they are carrying that they have never articulated. Each of the feelings is associated with something they tell themselves about themselves, something negative about their ability to grasp science. The leaders help the group as it begins to recognize the negative content of some of these self-statements. The student who said, "I feel just as if this were a real science quiz," is actually saying, "I must perform well in everything; I am always on trial, and I cannot afford to fail even once." Those who said they wished they'd studied more carefully are usually telling themselves, "Now I'll get just desserts: I'll be shown up for how stupid and lazy I really am." And those who apparently didn't take the quiz seriously, those who said, "It doesn't really count," are actually saying something they tell themselves regularly; namely, "I can't be expected to know this material, for the following reasons..."

These are only a few of the examples of the negative messages that students give themselves. They tend to fall into certain categories: those that deal specifically with science and those that are more general but tend to be exacerbated by the students' contact with science. The first two examples, "I must perform" and "I'll be shown up," are of the general kind, but they come out in connection with science. Students in our Clinic simply do not feel anxious about their performance in, say, literature or history. They think they can do well, and they generally do. When they do not, they seem able to take it in stride and resolve to do better the next time. What this means is that they haven't got a lot of ego tied up in how well they do in these courses. But a failure in science seems to tell them something terrible about themselves: They're stupid, or lazy, or just not "destined" to succeed.

The third example, "I can't be expected to know this because...", deals specifically with the student's perception of the nature of science. Unlike literature, it is not part of everyday life; you "can't be expected" to know anything you weren't taught in the classroom. Even though the readings that we assign for the quiz are fairly self-contained, the students tend not to see them as such. Those who are studying biology feel pressure to get the answers dealing with the article on radiation but excuse themselves from dealing with the others, which have to do with physics and astronomy.

The students' reactions to the answers are also very revealing. The first observation is that they consistently underestimate how well they did. Even if, just as an experiment, the scientist tells them that there were, say, two As, four Bs, three Cs, and one D, no student claims credit for either of the As; they virtually vie for the D. What the students have done is to relinquish their powers to assess their own performance. They have rather clear notions of how well they have done on a history test, but they have abdicated the responsibility for making a similar judgment about how they do in science. This heightens the aura of the mysterious and the unattainable that surrounds the study of science and is a strong contributing factor to science anxiety.

We have only touched on a few of the things students tell themselves in the course of taking the little quiz in the Clinic. Furthermore, not all the negative self-statements come out in connection with the quiz; if they did, we might suspect that we were dealing with test anxiety, which we are not.

The scientist then turns the discussion to one of science skills; in particular, the skills necessary to read science materials. The first thing the students are told is that science must be read slowly, with pen or pencil in hand; it should not be read at the rate we read nonscience books. This is a revelation to the students. They have, in trying to read it fast, been failing to grasp the content. This has led them to blame themselves since no one has ever told them how to read science. When the scientist tells them that even scientists read science slowly, that this is the right way and the only way to do it, the students have not only learned a new skill, but removed a source of anxiety. Each sentence and each equation, the scientist points out, are packed with information, and packed in a different way than literature or history. Each statement must be understood before you can go on to the next one, because the train of ideas builds one upon the other. And if it becomes necessary to stop and work out an example or two to get the author's idea, so be it. That's the way professional scientists do it, and that's the way that students must do it too. This means that reading a chapter of chemistry is going to take a good deal longer than reading a chapter of history; the students must learn to allot their time accordingly.

Does this mean that science is "harder" than history or that scientists are smarter? By no means. The good historian probably reads many books about the same material to acquire a number of points of view; the practicing scientist is able to read fewer, since "point of view" is not in principle germane to understanding. But the student has in general one history text and one science text, and the science text is going to take longer to read. The history student gets a taste of what historians actually do by writing term papers -- individual research efforts that require reading several sources. The science student gets a taste of what scientists do by working problems based on the text material.

The Clinic scientist then goes on to discuss the differences between the various readings. In the examples we use in our Clinic, the first article is a self-contained essay on radioactive fallout used on a standardized test of reading for content. It is somewhat easier than what one would find in, say, a college text in biology. The second is an excerpt from an article on black holes taken from the pages of Scientific American. This requires a good deal more concentration from the reader, since it deals with a variety of unfamiliar concepts. What it has going for it is the intrinsically interesting subject matter and lucid writing style.

The third article is taken straight from a freshman physics textbook for nonmajors. It deals with the concept of illuminance. It is not entirely self contained, although it requires relatively little prior knowledge, and it demands great attention from the reader. It presents the subject matter in standard textbook fashion, with a minimum of superfluous words and with equations to supplement and summarize the concept being presented. It is a little dull. It tends to make the students quite uncomfortable. This is the article they see as "real science."

The fourth article, about the life of Albert Einstein, is actually biography, not science, and presents no problem to the students -- except that some of them tend to tune out even the minimal science content in the article, as evidenced by their failure to state Einstein's contributions on the quiz.

The lessons to be learned from this discussion of science reading skills are that science in general cannot be read like literature, but that science, like literature, includes various sorts of materials, and that the easiest readings are not always the most rewarding. The black holes article, for example, is fairly difficult reading, but it is generally well liked.

One more question the scientist raises is that of quality of science writing. Certainly students make judgments about quality of literature. Why not make the same sorts of judgments about the science articles? They are not all of the same quality. This notion is a new one for the Clinic group. Science writing is treated as gospel: to be studied and learned. Any difficulty that the students encounter with the material they ascribe to their own inability to comprehend science. It almost never occurs to them that the author may have been unclear or even incorrect. Once more, they have relinquished their powers of judgment because of their distorted view of science. We ask the students to make an appraisal of the relative quality of the science writing in the articles we have assigned and encourage them to do the same with all the science reading they do from then on.

This completes the science skills part of the second Clinic session. The psychologist then turns to the restructuring of negative self-statements. The students have each brought in a number of negative statements they have heard in their heads in the preceding week. These are presented to the group and summarized by the psychologist and the scientist working together. It is essential that the scientist play a large role in this process, even though it may look superficially like the psychologist's domain. The reason is that the students' statements are often science-specific, but vague, and the scientist must draw on his or her own experiences and interpret for the psychologist. For example, a student might say, "I'm afraid to take a chemistry course in college because I remember in high school chemistry the practicum used to terrify me, especially in the unknown analysis." This is gibberish to anyone who has not studied chemistry. First of all, what is the nature of a practicum? What actually goes on? The student, with the help of the scientist, is encouraged to describe in detail what is actually involved: "The practicum is a test of how well we can do in the chemistry lab. We are given a lab bench, a bunch of chemicals, and one unknown chemical. We are then required to figure out what the unknown chemical is by mixing a little of it with the various other, known chemicals." Now things are a little clearer. "Unknown analysis" is not an adjective preceding a noun, but rather shorthand for "analysis of an unknown chemical." The actual lab situation is now something that the psychologist and the other group members can visualize. The scientist has acted as an interpreter for the student.

As each student presents his or her self-statement, the others respond in much the way they did to the science autobiographies. Many of the students have similar or related self-statements. It is supportive for the speaker to hear that others share the same feelings. One student says, "Every time someone asks a question in chemistry, I say to myself, "Why don't you know enough to ask such an intelligent question?" Another volunteers, "Every time I see a classmate studying biology in the library, I think everyone in the class understands the material but me." A third states, "When I was trying to work my physics problems this week, I heard myself saying, 'You'll never understand this stuff!'" A fourth student, a psychology major taking an elective course in science, reports that she has heard herself say, "I don't have the right kind of mind for this; how can I be expected to know it?"

Once the students have presented their statements and compared notes, so to speak, the psychologist introduces the approach that will be used to attack these statements and restructure them into more constructive self-talk. The model we use is the R.E.T. model of Albert Ellis.[2] The initials stand for "rational emotive therapy." The approach is just what it sounds like: We attempt to change the anxiety feelings by changing the reasoning of the statements that the students tell themselves. More rational self-talk leads to better feelings.

The psychologist explains the process by which the students make themselves anxious in terms of three steps:

137

    A. The stimulus or event.
    B. The self-statement that accompanies the stimulus.
    C. The feeling of anxiety that follows the self-statement.

The model seems extraordinarily simple when stated this way. The difficulty lies in the fact that the students are not conscious of step B. This is why we do preliminary work in identifying negative self-statements. An anxious student before coming to the Clinic experiences only steps A and C consciously, and erroneously concludes that A causes C: "Whenever I sit down to take a science test (A), my mind goes blank (C)." But it is the unconscious self-statement (B) that the student makes at the outset of the test that actually causes the anxiety. It is this self-statement that must be restructured. The actual sequence of events for the student who freezes on a science exam is something like this: "Whenever I sit down to take a science test (A), I tell myself, 'Science is too hard for you' (B), and my mind goes blank (C)." Or perhaps something like, "Whenever I sit down to take a science test, I say to myself, 'If you don't do well on this one, you're going to have to do very well on the next one. And if you don't, then there's a good chance you won't do well on the final. And if you don't, then you'll get a low grade in the course, which will bring your entire grade point average down. And that means that you're going to have to do extremely well in other courses, especially science courses, or your chances of going to med school or dental school or graduate school are nil.' And then my mind goes blank!" Now all this negative self-talk goes on unconsciously and takes a fraction of a second; all the unsuspecting student feels is a panic response to a science stimulus.

When the students have understood the outline of the R.E.T. model, they are given a new homework assignment: to write down examples of the whole sequence, as they experience it in the coming week -- the science stimulus, the anxiety response, and, as they become more conscious of them, the intervening negative self-statement(s).

In the final half hour of the session, we begin relaxation training. The students make themselves as comfortable as possible, either on their chairs or prone on the floor, and a standard relaxation tape is played. This tape instructs them how to tense and relax various muscle groups in their bodies. All parts of the body are dealt with so the students may begin to see where they express their tensions most. The group leaders monitor the students as they attempt the relaxation exercises and note whether they are actually able to relax. This relaxation procedure takes about a half hour the first time it is done. The students are then asked to practice by themselves at least 10 minutes a day every day. As they become more adept, it takes less and less time for them to relax on their own command, especially since they can focus on the muscle groups where they experience the most tension. We use a combination of relaxation tapes that we have made and some developed by Thomas H. Budzynski.[3]

After the relaxation training has been completed in this second session of the Clinic, the group members discuss how they felt, which muscles were easiest to relax, and whether they could get into the experience. Like cognitive restructuring, relaxation sounds simple on paper, but it takes some practice and commitment to really make it effective.

As the second session concludes, the group leaders collect the individual science anxiety hierarchies, which they will then mold into a composite hierarchy for the following weeks. The students are given new homework assignments: to write down one or more of the stimulus/self-statement/anxiety sequences they experience during this week and to practice the relaxation exercises daily.

Session 3

We begin the third session with a brief discussion of how the homework assignments went during the week and then move on to the so-called A-B-C sequence: stimulus/negative self-statement/anxiety response. Each of the students presents at least one of these sequences, preferably one that has occurred in the last week. The students have by this time become fairly good at recognizing when they are feeling anxious and identifying the situation in which the anxiety occurs. The trick is to help them get at the intervening negative self-statement. In fact, this activity will be continued in all the sessions for the remainder of the Clinic. As each of the students presents his or her sequence, the rest of the group provides support and feedback in the usual way: by identifying those situations that they share in common. The group leaders help clarify and summarize the students' descriptions and, most important, help identify the negative self-statements hidden therein.

Here are some typical sequences the students might present:

1. Even though I studied all week for my chemistry exam, as I'm walking to the exam, I feel anxious, and break out in a cold sweat. I tell myself, "You should have studied more."

2. When I try to work some homework problems in physics, I notice I'm having trouble concentrating. I'm thinking about everything except the task at hand. I realize that this is my way of feeling anxious. And what I tell myself is, "If you can't work these problems, then you'll fail the next exam, and the following exams will depend on your knowing the earlier material. So you may flunk the course. Then what will you ever become?"

3. Whenever I come to a point in my biology text where I don't understand something, I begin to feel scared. What I'm telling myself is, "Everyone else in the class probably understands this. And no one but you is feeling anxious."

Notice that each sequence contains the event that stimulates the anxiety, a description of the way the anxiety is felt, and a negative self-statement that causes the anxiety response.

When the students have expressed their negative self-statements, the group leaders point out that these stem from a series of basically irrational beliefs that the students have about themselves and about how the world works. These beliefs fall into a limited number of categories; three of the most common for science-anxious people are:[4]

1. Worrying helps. The student who always says "I should have studied more" is invoking worry as a kind of magical formula to help him or her perform better. This kind of ritual worrying is very popular among students. We all know the type of student who perennially complains about not having studied enough.   This may be harmless -- unless it is accompanied by anxiety in oneself or it provokes anxiety in others. Actually, the origins of primitive religion lie in just this kind of worry ritual. The anthropologist Bronislaw Malinowski describes the plight of the cave-dwellers who, having done everything possible to protect themselves from the ravages of nature and still mindful of their own weakness, will then pray to an icon for lack of any further logical steps to take.[5]   The student who worries after having spent a lot of time studying is doing the same thing. The difference is that the worrying itself does not help.   This is the obviously irrational part of the student's belief system. On the contrary, the worry leads to anxiety on the test, thus distracting the student and leading to a worse performance than would otherwise occur. The other pernicious aspect of the worry ritual is that it reinforces the student's already irrational belief that he or she is incapable of assessing when the science material has actually been learned.

2. A second common category is the belief that one's self-worth is determined by one's performance. The student who believes that being unable to solve the physics homework problems means not being able to amount to anything is a victim of this irrational belief. (The process known as "catastrophizing": "If I don't do well on this, then I won't do well on the next thing, etc." often accompanies the irrational connection between performance and self-worth.) This connection is difficult to counter, since society reinforces it at every turn. Why else would we buy beer because some football player drinks it?

3. The third common category can be described by the question, "Other people can do it; why can't I?" The third student subscribes to this belief.   He or she thinks that other people do not experience the same stresses and uncertainties, that they have some magical coping skills that allow them to avoid anxiety.

The task of the group for the rest of the Clinic sessions is to identify and counter these maladaptive beliefs as they are expressed in negative self-statements. Certain simple steps can help a lot in this task. The first is to look for semantic clues that indicate negative self-talk. Words such as "should" or "must" are tipoffs; so are generalizations from specific situations, such as connecting the physics homework problem with what the student will "amount to." A second simple step is to pick apart the self-statement or belief and decide which parts are rational and which are irrational. Usually there is some small rational "kernel of truth," but it is accompanied by a lot of irrational trimming. For example, the belief that "everyone in the class but me understands this concept" has as its kernel of truth the fact that some people in the class probably do understand the concept and have grasped it without too much trouble. After all, people have a range of talents. But the belief that everyone else has grasped the concept is patently irrational and must be confronted as such. (The technique for confrontation is taught in Session 4.)

The science skills part of Session 3 begins with the assignment of Problems 1 and 2 from Chapter 4 (page 76).

The students are given five minutes to work these problems and are instructed to listen for negative self-statements as they do so. They are told to ask themselves:

Is my attention on the problem or am I distracted?

Am I focused on myself? What am I experiencing?

If I don't finish the problems, what are my self-statements and feelings?

If I do finish the problems, what are my self-statements and feelings?

After the five minutes, the scientist works out the problems on the blackboard. The purpose of this exercise is twofold: to get at some of the self-statements about solving word problems and to teach the students the skill of solving these problems.

Many more students are able to solve Problem 2 than Problem 1. Why is this the case? It is obvious once they have been worked out that they are really one and the same. Yet the students tend to perceive them as very different. Some of the students read the first one, decide quickly that it is beyond them, go on to the second one, solve it, and then realize that the first is very similar. The rest never recognize the similarity. What is intruding in both cases is a series of self-statements about science. The most common one goes something like, "Who am I to be calculating the age of the universe?" Another is, "I can't be expected to know this; I haven't had astronomy." Still another is, "There's not enough information here. Something has been left out. I need to know more in order to do this." All these distract the students from the task at hand. Even those who get the solution usually do so after giving up and going on to the second problem and then seeing the similarity. The irrational belief underlying the negative self-statements is that there is something magical about science, that some people can do it and others can't, that one has to see the whole forest, rather than learn by blazing a trail from tree to tree. But the solution to the problem, its very similarity to the problem of traveling from Chicago to Detroit, belies these beliefs. What the science-anxious students have done is convince themselves that there are certain problems they can solve, such as math word problems that they mastered in earlier grades, and certain problems they can't solve, such as those having to do with science that they are approaching now, in college.

Now there is, as we have said, usually a kernel of truth to each irrational belief, and this is true for the universe problem. The fact is that the description of the universe in Problem 1 is a simplified model: The galaxies are assumed to move at constant speeds (they don't quite), and we also assume that the same sort of description of the universe applies today as when it started, it's just bigger. So the anxious student, or any student for that matter, might become bogged down in worrying about the departure of the model from reality. However, this is also true for Problem 2! The car doesn't always move at constant speed; it must start up in Chicago and slow down and stop in Detroit. The road from Chicago to Detroit is not straight. Nor is it exactly 300 miles. Yet this never bothers the students because they are used to idealized models of reality in areas in which they feel comfortable, such as math word problems. It is in science that they have come to believe that they cannot believe the models and therefore

Reprinted with permission from *Chemistry* (now *SciQuest*), pp. 6-9, October 1978. Copyright 1978 by American Chemical Society.

cannot solve the problems. Part of the task of the Clinic scientist is to help the student feel comfortable with the models of nature, while keeping in mind that they are only models. This is, of course, also the task of the students' science teachers, but for the science-anxious student a good deal more convincing is needed.

An interesting question regarding these two problems was asked by Professor Herbert Gottleib of Queensborough Community College: Is the scientific notation of Problem 1 also a source of anxiety? For the student who suffers from math anxiety as well as science anxiety, the answer is, probably. Dr. Gottleib suggests two additional word problems to determine if this is the case.[6]

Problem 3.

Convertible securities are commonly transmitted by bonded carriers between major industrial complexes and financial centers. Calculate the transport delay interval if the distance between two such centers is 5.00×10^2 kilometers and the carrier speedometer indicates a constant speed of 2.50×10^1 kilometers per hour.

Problem 4.

A nuclear reactor produces slow neutrons. How long does it take for a neutron with a speed of 60 feet per second to make the trip from the reactor to a detector 300 feet distant?

The solutions are as follows:

Solution 3

What We Know	What We Want	What Relates Them
$d = 5.00 \times 10^2$ km	$t = ?$	
$v = 2.50 \times 10^1$ km per hr		$d = vt$

$$t = \frac{d}{v} = \frac{5.00 \times 10^2}{2.50 \times 10^1} = 2.00 \times 10^1 \text{ hr}$$

The answer is 2.00×10^1 or 20 hours.

Solution 4

What We Know	What We Want	What Relates Them
v = 60 ft per sec	t = ?	d = vt
d = 300 ft		

$$t = \frac{d}{v} = \frac{300}{60} = 5\,\text{sec}$$

The answer is 5 seconds.

What we are dealing with in all four problems is an example of the transition from concrete to formal reasoning: That is what modeling nature is all about. As we discussed earlier, this transition is not automatic and is a source of science anxiety. Here, in the solution of these science problems, we have found not only a vehicle for teaching science problem-solving skills and a means for getting at some irrational beliefs, but also an opportunity to assist the students in making the transition from the concrete to the formal. One last remark: We choose problems that do not require prior knowledge of any particular science so that no student has an advantage.

The final part of the session begins with relaxation practice. A relaxation tape is played, and the students practice relaxing various muscle groups. When they are relaxed, the desensitization process begins. We use the model developed by Donald Meichenbaum.[7]

The first thing the group leaders do is to have the students imagine a personally relaxing scene: lying on the beach, for example. The students are asked not to view the scene from outside, but to place themselves in the scene and imagine all the details: how the sand feels on their body, the sound of the surf, the smell of the sea air. This is an important bit of imagery, since they will be asked to return to this scene each time they feel anxious.

Once the students have successfully imagined the peaceful scene, the desensitization process begins. The first item of the composite hierarchy is presented by the psychologist. This is the one that causes the lowest amount of discernible anxiety. The students are asked to imagine the scene and raise a finger if they are feeling anxious. The scene is repeated, until no anxiety is felt by any of them. This is accomplished by moving back and forth between the science scene and the peaceful scene. Once the first item has been overcome, the psychologist moves on to the next one. No more than three or four items are presented each week.

At the conclusion of this first desensitization session, the group discusses how it felt and the leaders make various suggestions for facilitating relaxation, including things to say to oneself and methods of breathing. Then the group is dismissed, with the usual homework: daily practice of the relaxation exercise and further recording of A-B-C sequences.

Session 4

This session begins like the others, with a brief discussion of how the week's homework has gone: the relaxation practice and the A-B-C sequences. The group leaders help focus on the irrational beliefs that underlie the negative self-statements. By this point in the Clinic, the students are really beginning to feel the effects of both the relaxation and the cognitive restructuring and are reporting positive results. The group leaders must be aware that the students will all try to give positive reports, to show they are indeed making progress. This unconscious "pleasing the teacher" has to be confronted, so that whatever problems the students are actually having are brought out in the open. Thus the group leaders caution the students not to give more positive feedback than is actually the case.

The science skills addressed in this session have to do with getting the most out of a science lecture. The first step is to prepare for the lecture. Preparation entails reading the chapter to be covered before coming to class. This seems like an obvious sort of study skill, but the reality is that in many courses students can get the gist of a lecture and then go home and read the text. This is in general not possible in science courses. The lecture material, like the reading material, is compressed into "scientific shorthand" -- briefly stated logical progressions from one concept to another, frequently accompanied by equations. It is the rare student indeed who can come in "cold" and get anything out of this process. In fact, as the scientist in the Clinic points out, the text must ideally be read three times: once before the material is covered in class, once during the time that the material is being covered, and once after the material has been covered. A primary function of the science lecturer is to clear up obscure parts of the text; therefore, the student has come to class already knowing which parts are unclear to him or her.

The next idea that the scientist must get across to the group is that the way science is done and the way science is presented are two different things. The lecture is deceptive: It lays out in logical progression the results of the creative research, clever guesses, and missteps of a host of scientists. Students much learn that while they can appreciate the elegance of the presentation, they should not attempt to emulate it when they are actually doing science, because that would only lead to more negative self-talk about how lacking in intelligence they are.

What about note-taking? What can you expect to get out of it? What if the professor suddenly says something you don't understand? Should you stop taking notes and think about it for a few minutes? Or should you continue to write, even though you don't understand what was said? The Clinic scientist points out that a good rule is to mark the place in your notes where you lost the point and then to continue taking notes. It does not mean that you have lost the whole train of the lecture. In fact, even where you apparently have missed nothing in a lecture, there are ideas presented that you will only understand when you look over your notes at home. Note-taking is more than simply recording a lecture. It involves making judgments about what the instructor considers important; it illuminates the text material. Listening to a lecture includes picking up the clues about what's important from how the instructor says things, from what is stressed and what is not. When you look at your notes at home, especially with the textbook at hand, you can often find out what the point was you missed, as long as you know where it appeared in the lecture.

When is it appropriate to ask a question in class? Ideally the answer should be: "As often as necessary, provided you have come in well prepared for the lecture." In practice, it varies from one teacher to another. Some are willing to answer a lot of questions. Others less so. The science-anxious student usually asks no questions in class, out of fear of ridicule from the instructor or from other students. As far as the other students are concerned, this fear is generally groundless. As far as the instructor is concerned, according to students we have talked to, this is unfortunately not the case. Some instructors respond with sarcasm to an earnest question, if they think the student should know the answer. What can the student do, especially if he or she is already anxious about science? The first thing is to separate one's ego from the instructor's response. Often, a sarcastic reply from a teacher gives the student the message that he or she is stupid. While the instructor's sarcasm is not excusable, it is important for the students to recognize the fault is not their own. It is also important for them to realize that there are some situations where they are powerless to effect change. They cannot force the instructor to be less sarcastic. They may even wind up asking fewer questions than necessary and have to find a way of getting the information without the instructor's help. But they must recognize that this situation is not one in which they are in any way at fault.

It is clear by now that there is no rigid separation between science skills and cognitive restructuring. Each of the science related situations involves some skills and some self-statements. The psychologist and the scientist work together to deal with the situation as an organic whole. The Science Anxiety Clinic is, after all, more than simply a class in remedial science skills.

The group now launches a direct attack on the negative self-statements of its members. Each student has brought in at least one A-B-C sequence: science stimulus, negative self-statement with its attendant irrational belief system, and anxiety response. The group focuses on the irrational belief of each of its members and helps that person devise an appropriate "coping statement." How is this accomplished? The first step is to identify clearly the negative self-statement and the belief behind it. Then the group searches for an objective statement that challenges the belief. This challenge must be carefully constructed. It cannot simply discount the belief, and it cannot be a pep talk. It must validate the objective conditions in which the student finds him or herself and then move on to the irrational aspects of the belief associated with the stimulus. Here is an example:

A student says:

"This week, as I was trying to study for a chemistry test, I noticed I couldn't concentrate. I was biting my nails, and I felt generally nervous. I thought about what I might be saying to myself, and I realized I was telling myself that I'll never be able to learn chemistry."

Here the student has given the stimulus (studying for the chemistry test), the anxiety response (nail-biting and inability to concentrate), and the negative self-statement ("I'll never learn chemistry"). The other group members and the leaders then ask the student questions:

"How did you decide you'll *never* learn chemistry?"

The student replies, "Well, I did poorly on my last test, so why should this one be any better?"

"How does your performance on the last test determine your performance on the next one?"

"Well, it shows that I'm no good in chemistry."

"Did you learn anything from the errors on your last test?"

"Yes."

"Then why do you predict no improvement on the next one?"

"I just feel that way."

"If you had done well on the last test, would you then be confident that you will do well on this one?"

"No, not really."

"Then why does a bad performance in the past mean a bad performance in the future, but the same is not true for a good performance?"

"Because the good performance would have been lucky; the bad performance is really me."

Now some of the irrational beliefs are beginning to come out. The group presses on:

"Isn't the word *never* a bit final?"

"How can you tell that you'll never do something in the future?"

"How can you conclude on the basis of one exam what your entire future will be with chemistry?"

Eventually the student is forced to come to terms with the irrationality of his beliefs. Notice that there has been no discounting or pep talks, such as "We know you can do well, don't worry about it." Such statements are worse than useless; they make the student feel more isolated and misunderstood.

Once the student has confronted the irrational beliefs, the group helps him work out some coping statements: rational self-statements that will challenge the negative beliefs and thus help the student function less anxiously and more productively. Such statements for this chemistry anxious student might sound like:

"There is no point in worrying about the last exam: It will not affect this one. The main thing is to focus on the task of studying. One exam is just that: one exam. It does not determine my entire future. It is irrational to believe that only my bad performances reflect the true me."

This restructuring process doesn't happen instantly. But in the course of this session and the next, each student has an opportunity to present at least one negative belief and have the group help challenge it. The other advantage of this exercise is that the students see how much simpler it is to focus on someone else's irrational beliefs than on one's own. Since many of the students have similar beliefs, it is a revelation to them to see their own negative self-statements held up to the cold light of day not only by themselves but usually by at least one other group member.

At the conclusion of this part of the session, the group leaders briefly discuss with the students how the process felt. They ask the students to continue recording A-B-C sequences and to try to initiate some coping statements as homework.

The final part of the session is a continuation of the relaxation and desensitization procedure. The relaxation usually takes a good deal less time by this point in the Clinic. About 15 minutes should be sufficient. The three or four items that are presented for desensitization in this session are medium-anxiety items and will therefore take somewhat longer than the low-anxiety items of the previous week. It is very important to make sure the students can very clearly visualize their "peaceful scene" with themselves in it, because they will be returning to it more frequently during this desensitization session. As the level of anxiety increases, more students will have trouble with certain scenes; that is, they will have difficulty getting beyond these scenes. The psychologist will then introduce other coping mechanisms directly into the desensitization: "If you are still feeling anxious, return to your peaceful scene," or "If you feel anxious now, repeat to yourself the words, 'calm, relax, calm,'" or, "If you are still feeling anxious, take one deep breath, hold it, and then expel it in five short puffs."

After three or four of the items have been completed, and all the students have been desensitized to them, the fourth session concludes. The students are asked to continue their relaxation practice at least 10 minutes each day. Some of the students may report at this time that they tend to fall asleep during their relaxation practice. The group leaders suggest that they carry out the practice in a position and at a time that would help them stay awake. Sitting in a chair in the middle of the day is better than practicing in bed in pajamas at midnight.

This point in the Clinic is a good place to deal with any resistance that group members may be experiencing with relaxation and desensitization. Not everyone can get into this procedure very easily. To some it seems a little silly and magical. The group leaders can provide the students with information about the success of the technique and even give them a brief list of references in the literature if they wish to read more about it.

This is also a good point at which to check on the students' sleep habits. Those who fall asleep during the relaxation practice in and out of the Clinic tend to be those students who are unable to tell themselves when to stop studying. They will frequently try to stay up all night before an exam, especially a science exam. It is important that the scientist and the psychologist point out that such habits hinder their studying, keeping them tired and more anxious. Also at issue here is the negative statement "No matter how much I study, it won't be enough," which must be challenged at future Clinic sessions.

Session 5

The fifth session continues much of what was begun in the earlier sessions: science word problems, negative statements and coping statements, relaxation and desensitization. The group leaders begin by asking the students to hand in their homework: the A-B-C sequences and the coping statements that they have developed to combat their negative self-statements. In addition to discussing these in the Clinic, the group leaders will study them during the week and make suggestions in the following session about how they may be better formulated. As the students present their own coping statements in the group, the leaders watch for the kind of "pep talks and discounts that hinder the development of useful coping statements. A student whose irrational belief is that worry helps, for example, may think that a good coping statement is "What are you worried about? Stop worrying!" This is not so. Such self-statements have just the opposite of the desired effect. They make the student feel even more peculiar than before, because they imply that there is something crazy about the worrying in the first place, without specifically focusing on the cause of the worrying. A much better set of coping statements is: "It is OK to worry a little; after all, the chemistry course certainly counts for something. But too much worrying distracts from the task of learning the material. And it is irrational to think that the worrying itself will help your performance."

The group leaders can point out a number of general skills to help the students combat their anxieties. One is time structuring. Many students, especially those who cannot decide when they have studied "enough," will be in conflict about when to work and when to play. The net result is that they will do neither well. When they are sitting down with their biology homework, they will be thinking about how much they would like to be playing volleyball. This will distract them from studying and raise their anxiety about doing well in the course, especially since they are unable to decide when enough is enough in studying science. However, when they then go and play volleyball, they will be thinking that they should be studying biology! The whole purpose of the physical exercise is then lost; they are not relaxed or refreshed by it and they return to their studies more anxious than ever.

The group leaders can show the students how to structure their time to avoid this kind of conflict. What is needed is that time periods be set beforehand for studying and for play, and then be rigidly adhered to. In our experience, students will be more likely to stick to the study time than to the play time. When it comes time to relax, they will more often than not say to themselves, "I should really keep studying." While this sounds admirable, it is actually a negative self-statement: "I cannot ever afford to stop studying -- I can never study enough; If I stop, then I will do poorly." If they believe this, then they will make it come true on the next exam, by their own anxiety. There's no prophecy so powerful as a self-fulfilling prophecy.

Another technique the group leaders can teach the students is to transfer their own coping mechanisms for nonscience-related anxiety to science anxiety. For example, a student may be quite anxious about going to the dentist, yet somehow get through it. How? The student has developed some coping mechanism, some statement that counters the anxiety and allows the task of going to the dentist to be carried out. Whatever the student uses can probably be adapted to combatting science anxiety. Another good example of common nonscience anxiety is stage fright, the fear of getting up and talking or performing before an audience. The student who is able to do so has obviously conquered this anxiety and can adapt the coping mechanism to science anxiety.

The science skills part of Session 5 involves the solution of more word problems and focuses especially on inappropriate "fallback" procedures the students use when they are unsure of how to solve a science problem. Rather than try to attack the problem directly, they will tend to use some technique they learned in high school (usually for math), regardless of whether it has any relevance to the problem at hand. Since they have convinced themselves that the science problem is beyond them, they will use anything they can, just because it feels familiar. Here are two typical science problems that illustrate this:

Problem 1.

By the cultivation of microorganisms is meant the process of inducing them to grow. For most purposes of microbiology they are cultivated *in vitro*, i.e., in the test tube. Cultivation in vitro necessitates the preparation of substances that microorganisms can use as food. Such nutrient preparations are called culture media.

Assume certain microorganisms will grow on nutrient agar at a rate proportional to the number already there. If 30 microbes are formed per hour when 10,000 microbes are there, how many are formed when 50,000 microbes are there?

Problem 2

A planet travels in an ellipse around a sun. When it is traveling at a velocity V_1, its distance from the sun is R_1. When at velocity V_2 its distance from the sun is R_2. The product of V and R is constant. When R_1 is 10^8 miles, V_1 is 350 mi/hr. What is V_2 when R_2 is 2×10^8 miles?

The students are given five minutes to work the problems. As usual, they are also asked to focus on any anxiety they feel and any negative self-statements they hear themselves making. These are discussed at the conclusion of the five minutes and again after the answers have been given. Here are the solutions to the problems:

Solution 1

What We Know	What We Want	What Relates Them
number $N_1 = 10,000$ when growth rate R_1 $R_1 = 30$ per hour $N_2 = 50,000$	Rate R_2	R is proportional to N; this means $\dfrac{R_2}{R_1} = \dfrac{N_2}{N_1}$

Solution: $\dfrac{R_2}{30} = \dfrac{50,000}{10,000}$ $R_2 = 30 \times \dfrac{50,000}{10,000} = 150$ microbes per hour

Solution 2

What We Know	What We Want	What Relates Them
$V_1 = 350$ mi/hr $R_1 = 10^8$ miles $R_2 = 2 \times 10^8$ miles	V_2	$V \times R = $ constant; this means $V_1 \times R_1 = V_2 \times R_2$

Solution: $V_1 \times R_1 = V_2 \times R_2$

$$V_2 = V_1 \times \frac{R_2}{R_1} = \frac{350 \times 10^8}{2 \times 10^8}$$

$$\frac{350}{2} = 175 \text{ mi/hr}$$

We observe that there is an *inverse* relationship between velocity and distance from the sun. The farther out the planet is, the more slowly it is moving. This is contrasted with the situation in the first problem, in which there is a *direct* relationship between growth rate and number of microbes: the more microbes there are, the faster the growth rate.

Students tend to get the first problem right more often than the second. Why is this the case? It is not because they understand the first problem better, or that it is more simply stated than the second, because it certainly is not. The reason is the following: The science-anxious student is very likely not sure about how to do either problem. What he or she does recognize, however, is that three numbers are given, and a fourth is asked for. So the student falls back on a familiar skill from high school math: taking ratios. Now, as it turns out, this is the appropriate thing to do for Problem 1, but not for Problem 2, which depends on "inverse ratios": $V_1/V_2 = R_2/R_1$. This comes from the statement that the product of V and R is a constant. The scientist points this out to the students and then may go on to discuss other examples where the relationship between the variables in the problem is even more complex. The students' math skills are adequate to the task. What the scientist and the psychologist must help them realize is that they can "see through" the problem to which mathematical technique is required -- as long as they can keep their anxiety under control; that is, as long as they don't tell themselves beforehand that the problem is beyond them.

The discussion of science problems leads naturally into a discussion of how to approach science homework. Students tend to believe that if you work on a problem for about 10 or 15 minutes and you can't get it, then you'll never get it and you might as well give up. This is not so. First of all, 10 or 15 minutes is generally not enough. The first attempt at a problem should be allotted a half hour, perhaps more. Second, if the first try is fruitless, it is best to put the problem aside for a while, even for a day or two, and let your mind work unconsciously on it. This is the way practicing scientists describe how they solve problems, and it is the right way to go about the task. Often, after two or three apparently fruitless attempts, the student will get a flash of insight and solve the problem. This is a thrilling feeling, and it is one of the rewards for which scientists work. The student who experiences this really begins to understand the excitement of doing science. And, if it turns out that the student cannot solve the problem, he or she at least knows the point at which he or she is stuck and can ask the appropriate question when the instructor goes over the problems in class. This is much better than making a brief attempt at the problem and then waiting for the instructor to work it out. The instructor's solution will look deceptively simple and will lull the student into thinking the material is easy or, alternately, into a negative self-statement: "I must really be dumb not to have seen that." Neither is actually the case. As we have said a number of times, the way science is presented is not the way it is actually done. Only by doing the problem, i.e., by making several serious attempts at it, can the student get familiar with the process.

After the science skills part of the session is completed, the relaxation exercises are done and the group is desensitized to the next few items in the composite science anxiety hierarchy. For homework, the students will continue relaxation exercises every day and will practice imagining the scenes that made them anxious during the desensitization. That is, they will reinforce the desensitization experience at home. In addition, they will continue to write down A-B-C sequences and develop coping statements.

Session 6

This session begins with a discussion of the homework and moves on to further presentation of coping self-instructions. The group leaders assist the students in finding these instructions but do not generate them for the students. It is essential that the students themselves discover the coping instruction in order for it to have the desired effect. In fact, learning the appropriate coping statements is a lot like learning science: You only get it by doing it, not by watching the teacher present it to you. The group leaders remind the students to be aware of the physical signs of anxiety: These are the cues that a negative self-statement is being made and needs to be challenged with a coping instruction.

At this point in the Clinic, the students will already be coping efficiently with some of their negative self-talk but will probably have discovered some new negative statements during the week. The group leaders point out that this is natural, that the Clinic provides no miracle cures, and that the students will probably become conscious of other negative statements even after the Clinic is over. What the Clinic does is to teach them to cope with such statements and to be aware of when they are feeling anxious. These techniques, coupled with the science skills and the relaxation/desensitization give them enough weapons to combat whatever science anxiety they are still experiencing even after the Clinic is over.

Session 6 is also a good place to deal with laboratory anxiety. Some students may be able to do well in the lecture part of the course but may panic in a lab. This problem can be attacked in a number of ways. First, for those students already taking a laboratory course, it is useful to have them go into the lab and experience the anxiety with the attendant negative self-statement. This can be done during their regular lab period, if it doesn't take too long; otherwise, they can probably obtain access to the lab at some other time and focus on their feelings without cutting into the actual lab work time. The purpose of this is to have them already aware of their negative self-statements when they come to Session 6. During this session, the group may actually go into a science laboratory. As in the case of the science word problems, it is important to choose a laboratory situation that is realistic yet that does not give those students who have had, say, a chemistry course, an advantage over those who have not.

We have found that the first experiment in many basic physics labs fills the bill nicely. It involves the measurement of the weight and volume, and calculating the density, of objects of various materials. One needs no advance knowledge. The concepts are simple and can be explained on one sheet of paper. The students are given the sheet and a lab setup and are asked to make the measurements in about 15 minutes. They are also asked to focus on any anxiety they feel, as well as on any negative self-statements they hear. At the conclusion of the experiment, they are asked how the experience felt. The scientist then gives the correct answers, goes over the appropriate procedures to make the measurement, and once again asks the students how the experience felt. The scientist can, if necessary, outline a procedure for laboratory that is similar to that for problem-solving. Before students start the measurements, they should have prepared data sheets on which are contained what is known, what is to be measured, and what is to be calculated from the measured quantities. For example, in this experiment, weight and dimensions are measured and the density of the object is calculated from these measurements. What is known is the density of various materials, to which the experimentally obtained densities are compared.

The main negative self-statement associated with laboratory anxiety seems to be the same sort of "forest-trees" statement we saw earlier in the context of the problem about the age of the universe. The students do not believe you can get an experiment to work by getting each of the pieces of equipment to work one at a time. Rather, they think that you have to get everything working together at the first shot. In fact, the famous Murphy's Law is an expression of anxiety: "If anything can go wrong, it will!" The coping mechanism involves recognizing that the task will be accomplished by small, discrete steps.

One important idea to get across in the laboratory exercise is that the theory is only an idealized model of Nature, and that experiments are not exact; there is a natural range of error associated with the experiment. Many students are under the misapprehension that they are supposed to obtain exactly the same answer in each trial. They think they've done something wrong if they fail to do so. This leads to the negative self-statement, "It'll never work!" that lies at the heart of laboratory anxiety. Although good instructors do teach their students about experimental error and its causes, the science-anxious students never quite believe it and are therefore unable to distinguish between natural error and a mistake in doing the experiment. This problem must be dealt with in the context of the Science Anxiety Clinic, where the skill of doing a measurement is taught simultaneously with a discussion of the self-statements associated with doing the measurement.

The laboratory is also a place where the women in the group may feel particularly uncomfortable, since they're not "supposed" to be good in the lab. These types of self-statements must also be challenged if they come up.

As in an actual laboratory class, the students should be asked to work in pairs. The group leaders can then observe if one of the partners tends to hold back or avoid actual contact with the experiment. Partners can be switched in the middle of the exercise, so that people can work part-time with a person of their own gender and part-time with a person of the other gender. They should then be asked to describe their feelings in both cases, to see how this is related to things they tell themselves and to their anxiety. Just as the females may feel anxiety because they're not supposed to be good at it, the males may feel anxiety because they are supposed to.

For the final third of the session, the group continues with the relaxation and desensitization exercises. For homework, they continue these exercises at home, as well as listen for negative self-statements and develop coping statements.

Session 7

This is the final session. The desensitization hierarchy is completed, any remaining negative self-statements are dealt with, and any remaining questions about science skills are raised. The students are also given time to assess the Clinic experience and to say goodbye.

The session begins with a discussion of the homework. Any new A-B-C sequences and negative self-statements are brought up. The leaders and the other group members help the person with the new statement to begin to find a coping mechanism (as usual, most of the work on this is done by the person whose anxiety it is). The group leaders also raise the issue of how the students will use their new skills after the group is terminated. One interesting bit of feedback we often get from the students is that they are able to use many of these skills in nonscience-related situations. The relaxation techniques are applicable everywhere. So is the ability to focus on anxiety feelings and the negative self-statements behind them. Some students report that their grades in science have already improved. While this is nice to hear, the group leaders also point out that raising students' grade point averages is not the primary goal, except inasmuch as it is a consequence of lowered anxiety. The goal of the Clinic is to make students comfortable with science and cognizant of the importance of learning science, whether or not they get an A in the course. People have a natural range of talents in science just as in other areas; we simply want their performance to reflect their actual talent, rather than their anxiety.

How to take a science exam is the skill on which the group focuses in the last session. The first thing considered is how a science exam differs from other exams and how this difference may lead to anxiety. For the student whose primary expression of science anxiety is panic on the science exam, the differences between science and other exams is quite dramatic: "On other exams you can always write something down; on a science exam, even if you've studied, you may not be able to answer a single question!" This is certainly true for the student who "clutches" on science exams. There is no easy way to start working a quantitative word problem, like there is in writing an essay in history. Even if the answer to the essay question is wrong, the student is not anxious about it beforehand.

But the science word problem may look like a complete mystery. The student may even wonder if it has anything to do with the course material! One high-anxiety item that appears on the hierarchy in various forms can be stated (a little humorously) as, "I sit down and look at the exam, and then wonder if I'm in the right room!" So the first thing the group leaders do is to work on science exam skills. Then they focus on the negative self-statements and the anxiety.

The scientist first points out that the exam is usually like homework. In the case of physics and chemistry, this generally means quantitative word problems; in the case of biology, it may mean the description of biological organs and other structures and an analysis of their functions. In any case, the exam problems should be treated like the homework.

The scientist then goes on to explain what is the most efficient way to take an exam. This includes reading the entire exam through before starting, working the problems in your own order, from the one that seems easiest to the one that seems hardest, writing down something for each question (especially on short answer or multiple choice exams), and putting a mark next to those you are not sure of, so that you may return to them, time permitting. It also includes not changing all your answers in the last few minutes because they suddenly, in the heat of finishing, "look wrong." The scientist also discusses how the exam is different from the homework: Since it must test a wider range of concepts than any particular homework chapter, an exam question may draw from several different areas of the course. Thus the student has to look carefully at each question and determine what concepts are involved, whereas in the homework, this is more obvious.

This discussion leads naturally into self-statements about taking science exams. Those students who try to answer the exam questions in the order they are given are usually prone to the "catastrophizing" sort of self-statement:

"If I can't answer the first question, then how will I be able to do the second one? And if I can't do that, then my grade on this exam depends on my getting the third and fourth ones right. What if I don't get them either? Then I've flunked the exam and my grade for the course is going to be terrible. How will I ever get a good grade point average so I can go to dental school?"

In addition to the usual coping statements: "Focus on the task," "One exam doesn't determine your future," etc., the student must also be convinced to decide not to answer the questions in the given order but to assess which ones look easiest and to do them first.

Another self-statement popular among science-exam-anxious students is the one that trips them up even on the questions they can do: "If this looks easy, I must not understand the question." This is similar to the self-statements that "Everyone knows what's going on but me" and "If I ask a question in class, people will laugh, because I'm the only one who doesn't understand." On the exam, this self-talk leads to the student's reading something into the question that has not been asked for and thus solving the "wrong" problem. A good coping statement might be, "If this question looks easy, it's probable that I understand this part of the course pretty well."

Students whose anxieties are related to how they think others are doing are often very distracted on a test: "When I'm trying to work a problem, all I'm conscious of is that everyone around me is writing furiously, and I assume that they all understand the problems better than I do." Coping statements that emphasize the importance of focusing on the task and that test the realism in the students' perceptions -- "When I get the test back, has everyone really done better than I have?" -- are useful in challenging this type of anxiety.

We must emphasize once again that we are not dealing here with generalized anxiety about exams. These students rarely, if ever, experience anxiety on exams in nonscience courses. Therefore the discussion of exam anxiety must focus on its science-related content and must include con- sideration of the probability that the exam is only the ultimate expression of the anxiety that the students feel throughout their science courses.

Other items that the group discusses in relation to taking a science exam include: how to come physically prepared (a good night's sleep and no excess stimulants such as coffee beforehand); how to avoid anxiety producing situations before the exam (such as listening to your fellow students boast about how much they know or complain about how little they know, or ask you if you studied this or that topic), and ways to do some verbal and physical relaxation exercises just before the exam.

After the conclusion of the discussion of exams, any other situations that have not been touched on are raised by the students. These, of course, vary from group to group.

The scientist concludes the discussion of science skills with a brief statement about the applicability of science to everyday life, reemphasizing the importance of students having a comfortable relationship with science even if they have no intention of going into science-related fields. Then the group discusses the termination of the Clinic. People share their assess- ments of the Clinic, make recommendations, and share some of the feelings they have about the group's ending. It is also highly appropriate for the group leaders to share their feelings and assessments at this time.

The group concludes with a relaxation session, followed by completion of the desensitization hierarchy and a final brief group feedback on the desensitization. Students are then given individual appointments for 10-minute "postsession" interviews with the group leaders.

Postsession Interview

Within a week or so after the end of the Clinic, each student meets with the group leaders for a brief discussion of the Clinic and what the student has found useful or not useful. This is also a time when the leaders

Chapter 6

can give the students some feedback about their participation and about what still needs to be done to combat their science anxiety. Any data that the leaders may want to obtain for Clinic records or for researchers in science anxiety may be obtained from the student, with the usual protections of "informed consent" invoked. The group leaders let each student know when and where they will be available if the student runs into any further problems with science anxiety.

If, during the course of the Clinic, the psychologist decides that a particular student has problems outside the range of science anxiety and would benefit from other types of counseling offered by the Counseling Center, the postsession interview is the place where this should be suggested. As we have mentioned a number of times, science anxiety in some cases is an "early warning system" for other personal problems; these may be revealed in the course of the Clinic, and (if a student is willing) can be dealt with by other counselors after the Clinic is over. In only one case in our experience was a student admitted to the Clinic, and later asked to leave the group by mutual consent, since his problems not only went beyond science anxiety but were interfering with the functioning of the group.

The final task of all the group leaders is to meet regularly with the staff psychologist who oversees the Clinic and discuss the experience as a whole, comparing the functioning of the various groups and making recommendations for improvements in future Clinics.

[1]E. Jacobson, *Progressive Relaxation,* Chicago: University of Chicago Press, 1938.

[2]A. Ellis, *Reason and Emotion in Psychotherapy*, New York: Lyle Stuart Press, 1962.

[3]For example -- T.H. Budzynski, "Relaxation Training Program" (1974), available from Biomonitoring Applications, Inc., New York.

[4]F.C. Richardson and R.M. Suinn, "A Comparison of Traditional Systematic Desensitization, Accelerated Mass Desensitization, and Anxiety Management Training in the Treatment of Mathematics Anxiety," *Behavior Therapy* 4:212-218, 1973; "The Mathematics Anxiety Rating Scale," *Journal of Counselling Psychology* 19:551-554, 1972.

[5]B. Malinowski, *Magic, Science, and Religion and Other Essays*, New York: Doubleday, 1948.

[6]H. Gottlieb, letter in *The Physics Teacher*, XXI, 340-341, (May 1983).

[7]D. Meichenbaum, *Cognitive Behavior Modification: An Integrative Approach*, New York: Plenum Press, 1977.

7

Beyond Fear:
Science For A Better World

I should like to conclude this book by reconsidering some of the problems that we discussed in the previous chapters, especially problems about people's attitudes and ways to change them, so that future generations will not fear science. Let us therefore take one more look at the condition of science in our society today.

The Limitations of Science

A gambler was interested in finding an accurate way to predict which particular horse would win a race in a given field of horses. He hired a statistician, a geneticist, and a physicist to study this problem and gave them six months to come up with an answer. At the end of the six months, the statistician and the geneticist presented the gambler with their solutions. The former had studied the odds and made some reasonable recommendations, while the latter had studied the genealogy of race horses and made some other reasonable recommendations. The gambler, however, was not satisfied with reasonable recommendations. He wanted a foolproof answer. He therefore called on the physicist, who, bursting into the room with a sheaf of papers covered with equations, exclaimed excitedly, "I have solved your problem exactly -- for a spherical horse!"

This little joke illustrates one of the basic problems facing science in its interaction with society: Just how is science applicable to "real" problems? Science, as we saw earlier, makes models of reality. It simplifies the picture that nature presents in order to help us achieve understanding. Only after this has been accomplished does science say anything about the effects of the things left out of the model. In Chapter 2 we discussed the observations of Galileo, in which he stated that all objects fell at the same rate in the *absence of air resistance*. But air resistance is never really absent on earth; it turns out to be a small effect over short distances for objects such as rocks, which are compact, but a large effect for, say, feathers. So we must be careful about the cases our models do not cover.

The nineteenth and early twentieth century view of the power of science was unduly optimistic. Writers such as H. G. Wells and Jules Verne believed, as did many of their contemporaries, that all the world's problems would be solved by science and its application, technology. What we have seen, however, is that with the twentieth century nearly over, the achievements of science have been a mixed bag: control of disease, new labor-saving devices, rapid transportation -- and nuclear weapons, industrial pollution, carcinogenic chemicals. Much of modern fear of science stems from the false hopes and unfulfilled promises that have attended its development. If we are to move beyond fear, we must understand the limitations of science. What can it *not* do? Where do we set the limits? When is model-making inappropriate? As science becomes an increasingly important component of politics, we need to know these limits.

Let us approach the question of the limitations of science from the point of view of making models, since what to include and what to ignore in these models is clearly a source of anxiety for students learning science, as well as a source of concern for people trying to make political decisions about such things as nuclear energy based on models presented by "experts." We must first recognize that a scientist investigating a new area of Nature begins by hypothesizing a simplified model of the phenomenon under investigation. The initial choice of what simplifications to make is based on aesthetics and experience, but it is still a guess. (If we knew the answer in advance, it wouldn't be research!) The justification for the model only comes later, if it is able to explain the phenomenon. If it is not; i.e., if the data do not fit the model, then an alternate model must be chosen.

This is the way we attack problems in "pure" science. What happens when we try to apply this approach to actual problems of society, technology-based or otherwise? A number of things can go wrong. The first is, we may not have enough data to decide if our model of a phenomenon is adequate. Good examples of this are pollution and nuclear reactors. When new chemicals or new industrial techniques were introduced into society over the last century, there was little or no motivation to seek data on long-term effects. People believed the nineteenth century model of the world: Technology would free us, with virtually no negative consequences. Now we are faced with some of the negative consequences: poisoned air, possible carcinogens in our water supply, and so forth, and we are just beginning to amass data on their effects. So, the nineteenth century "scientific" model of how we interact with the earth is in rather poor agreement with our new data. The problem is, we cannot simply change our model -- we have to find a way to clean up our act. The consequences of our lack of data are much more severe for society than they are for a pure science problem.

In the case of nuclear reactors, so much has been written about the accident at Three Mile Island that I hesitate to add to the torrent of words --except to say that while there seems to have been sufficient data on how machines work, there was insufficient data on how human beings work in the presence of a bewildering array of machines. This area of study is fairly new; reactors are presently in operation with little attention paid to human factors. Here again, because no one even thought of amassing data on this problem, we are paying the price.

A second area in which science can be incorrectly used is in the decision of what constitutes legitimate areas of study. The extreme example is, of course, the horrible experiments the Nazis performed on concentration camp inmates. Unfortunately, this sort of criminally irresponsible misuse of science was not confined to them. The U.S. Army's studies of biochemical warfare led it to secretly disperse a supposedly harmless gas into the San Francisco Subway System, leading to a number of illnesses and at least one death. The CIA fed LSD to unsuspecting subjects, resulting in one or more fatalities. Obtaining the "informed consent" of subjects in psychological and medical experiments is still an area of controversy among scientific researchers. Laser research safety procedures vary from country to country. So do procedures for working with recombinant DNA. There are many other examples. The net result of public knowledge of these situations is a general mistrust of scientists, a fear that much, if not all, research is pernicious in outcome, and a general attitude of antiscience that harms all scientists equally, not only those who have lent their talents to questionable experiments or taken inadequate safety precautions.

A third way in which science is misused is in applying its powerful methods to fields where such methods may be fundamentally inapplicable. We have all heard the expression, "statistics can be made to show anything you want them to." What is true about statistics is true about some of the other tools of science. They can conceal a lack of information or pervert information simply by the detail in which the information is buried. The essence of scientific advance is reproducibility -- anyone's results should be reproducible by anyone else. In actuality, except in controversial cases, we tend to believe the results of others since they are usually arrived at through years of hard work with expensive equipment. Furthermore, one gets no credit for repeating someone else's experiments; all of us wish to do our own original work. The result is -- we must trust one another. When this trust is betrayed, either through malice or mistake, the whole scientific enterprise suffers, for wrong information enters the public domain. Should the public find out, the reputation of science is tarnished further.

The overuse of certain analytical procedures -- statistics, computer analyses, and the like -- in certain areas of the social sciences is a good example of innocent but misguided application of some of the tools of science. The data on learning in twins, reported by Cyril Burt earlier in this century, appear to be an example of falsification of results. Burt, whose

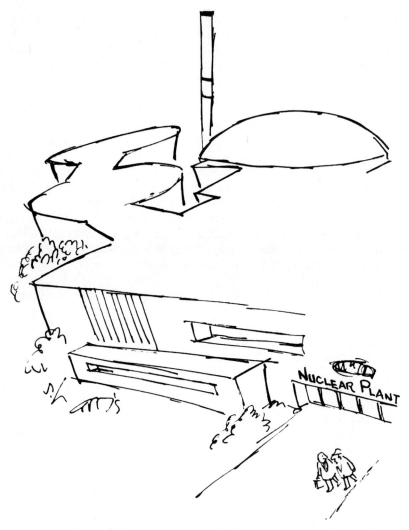

"Of course it's perfectly safe. Any accident would be in complete violation of the guidelines established by the Nuclear Regulatory Commission."

Copyright 1976 by *American Scientist* and Sidney Harris.

results have been used by later researchers to make claims about the different levels of intelligence of different races, may have made up his data and even invented coworkers to list in his publications.[1]

Among scientists, no less than between the public and the scientific community, trust is the cornerstone of progress. If people trust scientists, and scientists trust each other, then the scientific enterprise goes forward. If the tools of science are misused, applied to areas where they are not truly applicable, or worse, used for deception, then this trust evaporates.

The fourth way in which science can be misused is to apply it to political and social problems and to assume that society plays by the same rules as scientists. This can happen in a number of ways. The first is to treat societal conditions as if they are isolated problems. The nice thing about science is that the researcher can change each variable one at a time and see what happens. To assume we can do this with society is simply wrong. The "social engineering" approach to our problems has not been notably successful over the last two decades. It has led to further mistrust of science by the public. Let me give an actual example. There is a very eminent physicist who gives lectures around the country on how to deal with the energy crisis: whether nuclear energy is safe, whether we should use coal, whether we have sufficient oil reserves, and so forth. One argument he makes is in favor of strip-mining of coal. In answer to critics who fear that the land will be destroyed, this physicist points out that simply reinvesting a small amount of profits, some few millions of dollars, into reclamation and beautification of the land that has been mined will solve the problem. The error he makes is in assuming that the strip-mining corporations will act like impartial referees of scientific articles. In actuality, given a choice of spending a few million dollars to beautify the land as opposed to a few tens of thousands to influence legislators to allow strip-mining without reclamation, the companies will surely choose the latter course.

A similar argument can be made for reactor safety. Returning to the problem of Three Mile Island -- even if the machines and the human beings could in principle be made to operate safely, it was the desire to avoid payment of taxes that led to the decision to press this reactor into service before it was ready. So there is a certain naivete attendant to applying scientific models of operation to political conditions. To assume that logic and reason, rather than such things as greed and short-term perceived self-interest, are the dominant modes of operation is to misapply science in a myopic and dangerous way.

All the ways in which science has shown its limitations have by now entered the public consciousness. The beneficial aspect of this is that people have a healthy mistrust of the more grandiose schemes of those who claim to represent science. The negative aspect is that people have a general fear of science, a fear that they cannot always articulate but that is nevertheless there and that affects their ability to comprehend scientific ideas and to make important political decisions about such things as reactors, pollution, and genetic engineering.

The Successes of Science

Now let us ask, what are the great successes of science and how have they come about? What is it about science that inspires awe? Scientists have certainly developed a system that unlocks Nature's secrets.

And what just about everyone agrees is "progress," in the sense of increased understanding, is clearer in science than in any other endeavor. We know more about the universe, about biological organisms, about chemical and physical phenomena each day, and this progress is generally clear and easily seen. Sometimes it is seen as pure advance in knowledge -- the existence of black holes, the nature of the fundamental particles of which matter is made, the Big Bang theory of the origin of the universe are all ideas that have entered the arena of popular interest. Sometimes the public only sees the technological outcome of scientific advance: lasers, integrated circuits, and miniaturization of electronic devices, for example. In any case, scientists seem to have worked out a scheme for getting what they are seeking. If we try to discover just what this scheme is, we arrive at some surprising conclusions.

The pioneering study of how science actually works is Thomas Kuhn's *The Structure of Scientific Revolutions*.[2] Kuhn shows first of all that science, contrary to popular perception, has no *rules*. It has rather what are called "paradigms": ways of looking at nature that define which are the important problems that scientists try to solve. A scientific revolution is the replacement of one paradigm by another that explains more phenomena. It is the intellectual struggle between defenders of the old paradigm and proponents of the new that provides the creative tension leading to scientific advance, while at the same time keeping what is good about the old.

Essential to this process are two ideas that are important to us in our investigation of science anxiety. The first is that science has no fixed rules. It is the mistaken notion that it does that is the cause of anxiety among hosts of students. They are looking for rules in the form of "formulas," while science is truly learned by example, by *doing*. And in this sense, science looks more and more like art. You cannot learn to paint by writing down a series of rules. It is the same with science. By doing science, the student gets the idea of what science is without the necessity, and indeed without the possibility, of stating some set of rules. If there were rules, the process would be devoid of creativity.

The second feature of Kuhn's analysis is an idea about how science is *different* from everything else and how this difference leads to its great success in solving the problems it has set. No one is less concerned with receiving the approval of "outsiders" than is the scientist. The scientific community is insulated from external pressures to a greater degree than any other group, *and that is a main source of its strength*. Social scientists often justify their choice of research problem on the basis of possible benefit to society; artists and writers ultimately seek the approval of the public, but scientists choose problems and seek approval only on the basis of what is deemed important by their scientific peers. The greatest sin a scientist can commit (besides manufacturing false data) is to go to outside authority for approval of his or her work. That is not to say that we don't seek funds for our research from government and private agencies: We do. It does mean that the ultimate fruits of our researches stand or fall on the assessment of them by other scientists, not by state proclamation or popular support. Scientists who even announce the results of their researches to the press before these results have been assessed by referees of scientific journals can expect the censure of their colleagues.

This self-imposed isolation from societal pressures may seem elitist and undemocratic, but it is the core of scientific advance. Without it, science would be subject to the whims of the day rather than to the critical investigation that separates the wheat from the chaff.

You may ask at this point, haven't I completely contradicted much of what I have said in the rest of the book? Isn't the purpose of reducing science anxiety to bring nonscientists closer to the ideas of science and to scientists themselves? How then can scientists demand that they be insulated from society?

Utopia

The resolution of this paradox lies in our view of what the future would look like if we were successful in combatting science anxiety, and in eradicating it. Would this mean that everyone would then study science in school? Yes -- more than they do now, and without fear. Would everyone become a scientist? No -- people have different interests and inclinations. A few more people might become scientists, if science learning were anxiety free. Who became a scientist would no longer be restricted by sex and ethnic background. More women and minorities would enter the sciences. What then would be the function of a scientifically educated citizenry that was not, however, working in science? Just this: to assess claims of technology, to make intelligent political decisions based on scientific information, and to help professional scientists *by insulating them from societal and governmental interference.* Far from opposing the isolation of science from external pressure, a scientifically literate, nonanxious citizenry would recognize how crucially scientific advance is tied to this isolation.

Nor would scientists be tempted to "sell" their "product" on the basis of potential "practical devices" that might result. Science would be accepted on its own terms: as a creative endeavor, like art, music, and literature. It would not need some external justification, as it does now, to mollify a frightened and confused public. The job of the scientist would include explaining advances in science to the public in terms it could understand. Since science will for the foreseeable future depend on public funds, the scientists would have a strong motivation to do this. But they would no longer be running the obstacle course of science anxiety, and their ideas would find a ready audience. Why? Because the ideas themselves are intrinsically exciting.

The "mad scientist" stereotype would go the way of all stereotypes: into oblivion. The young boy or girl who was attracted to science would no longer have to ask, "Do I want to be like them?" The Two Cultures would be reconciled into the One Culture that they really are. People would have a

"Our sun is more than four billion years old and has already reached about half its life expectancy. It is now time to plan for the future of mankind, and a positive first step is the election of someone who is willing to face this vital problem . . . ."

clear idea of what science can and cannot do. The pseudosciences could no longer masquerade as science. "Experts" proposing one technology or another would really have to prove themselves to a critical citizenry. Scientists like me would no longer have to cringe internally when new acquaintances find out what we do, and say, as they almost invariably do (with a little note of pride in their voices), "Oh, I don't understand that at all!"

This Utopia I have just outlined is not exactly around the corner. The hundreds of letters I have received from science-anxious people, and from frustrated science teachers, attest to the magnitude of the task ahead of us. I hope that this book contributes to solving the problem of fear of science by showing science-anxious people how to cope with their fears and by showing science teachers and parents how to help students who are fearful of science. I also hope I have shown how people can set up Science Anxiety Clinics so that individuals need not remain isolated in their attempts to become scientifically literate. A universe is out there waiting -- not to be conquered, but to be understood -- willing to yield its secrets. Why shouldn't everyone have the key?

[1]For discussion of the controversy over the validity and honesty of Burt's data, see *Science* 194:916-919, 1976; 195:246-247, 1977.

[2]T. Kuhn, *The Structure of Scientific Revolutions*, Chicago: University of Chicago Press, 1962.

Appendix

Starting A Science Anxiety Clinic At A College Or University

In order to start a Clinic, there are three essential prerequisites: support of scientists and psychologists, available personnel, and a source of seed money. The scientific support can come from one or more of the science departments at the university. In our case, the Physics Department supported the founding of a Science Anxiety Clinic as a worthwhile endeavor, thus publicly stating that it recognized the problem of anxious students avoiding science. In addition, the science departments provide the science personnel (faculty members or graduate students) for the two-person teams that run the Clinic Groups.

The other half of the support can come from either a Psychology Department, a Counseling Center, or a Counseling Department in a School of Education. We have been fortunate to receive support from all three sources at Loyola. The major source of support has been the Loyola Counseling Center, which provided the location and the staff to start the Clinic. In addition, the Center finds interns in its programs to colead the clinic groups. Other coleaders have come from the Psychology and Counseling Departments. In addition, students who may be interested in doing research in science anxiety can be found in these departments.

It is essential, as I have remarked at several places in the book, that the coleaders not be both men (unless the Clinic is at a men's school). At least one of the leaders should be a woman and, ideally, a scientist, in order to combat sex stereotypes of role model. Obviously, due to the actual societal situation in which we find ourselves, the scientist tends to be male.

The Clinic is not a very costly project. It operates out of a room that can hold a dozen or so people and requires salary for the group leaders. For graduate students, this means the standard rate for a task requiring (for each group) about four hours per week for seven weeks. For faculty members, this generally means purchasing the equivalent of one course's time; i.e., salary for someone else to teach the course that the faculty member is relieved of in order to work in the Clinic. In our case, we received such funds from a Mellon Foundation Grant administered by Loyola, until we had proven the value of the Clinic, at which time it became part of the University's annual budget. Other possible sources of funding are those agencies within the federal and state governments that are interested in science education, and especially in educating those groups, such as women and some minorities, that are under-represented in the sciences.

The Clinic must be widely advertised. Students are sometimes oblivious to the services available on their own campus. If possible, Clinic workers should go into classrooms, *especially introductory science classrooms*, the first week of the semester and announce the Clinic. (Most of our science faculty have permitted us to do this and, in fact, are quite supportive of our efforts.) Enrollments in the Clinic are generally higher in the fall than in the spring.

Some prescreening of Clinic applicants is useful, to try to enroll only those whose main problem is science anxiety, rather than something else. Our simple science anxiety questionnaire (see the end of Chapter 3) is reasonable effective in doing this, although each institution's Counseling Center may wish to use its own screening procedures.

Attendance is crucial, and a clear commitment from students that they will be there except in case of emergency or illness, is essential.

Bibliography
Of Women And Science

The following deal with the problem of women and science. I am indebted to Dr. M. Grace and Professor J. Wittner of Loyola University and to the Biological Sciences Curriculum Study Project, Boulder, CO, for their help in assembling this Bibliography.

M. L. Aldrich, "Women in Science," *Signs* 4:126-135, Autumn 1978.

P. Broverman et al., "Sex-Role Stereotypes: A Current Appraisal," Unpublished manuscript, 1972.

S. Cotgrove, *The Science of Society*, London: George Allen Unwin, p. 315, 1978.

V. Crandall, "Sex Differences in Expectancy of Intellectual and Academic Reinforcement," in C. P. Smith, *Achievement Related Motives in Children*, New York: Russell Sage Foundation, pp. 11-45, 1969.

L. H. Fox, "Women and the Career Relevance of Mathematics and Science," *School Science and Mathematics* 76:347-353, 1970.

B. G. Glaser, *Organizational Careers*, Chicago: Aldine, 1967.

M. Guttentag and H. Bray, *Undoing Sex Stereotypes: Research and Resources for Educators*, New York: McGraw-Hill, 1976.

M. Horner, "Femininity and Successful Achievement: A Basic Inconsistency," in J. Bardwik et al., *Feminine Personality and Conflict*, Belmont, CA.: Brooks/Cole, 45-74, 1970; M. Horner "Towards an Understanding of Achievement-Related Conflicts in Women," *Journal of Social Issues* 28:157-176, 1972.

A. Kelly (Ed.), *Girls and Science*, Stockholm: Almqvist and Wiksell, 1978.

Cher King, "Career Expectations of High School Girls," *Region VIII General Assistance Center Newsletter*, March 1978.

A. Lantz and A.S. West. *An Impact Analysis of Sponsored Projects to Increase the Participation of Women in Careers in Science and Technology,Denver Research Institute*, University of Denver, 1977.

N.R.F. Maier and G.C. Cosselman, "Problem-Solving Ability as Factor in Selection of Major in College Study: Comparison of the Processes of 'Idea-Getting' and 'Making Essential Distinctions' in Males and Females," *Psychological Reports* 28:503-514, 1971.

A.S. Rossi, "Equality Between the Sexes: An Immodest Proposal," *Daedalus* 93:607-652, 1964.

A.S. Rossi, "Women in Science: Why So Few?" *Science* 148:1196-1202, 1965.

F. Scardina, *Sexism in Textbooks in Pittsburgh Public Schools, Grades K-5*, Pittsburgh: KNOW, Inc., 1972.

L. Sells, "California Women in Higher Education," available through Women's Center, Building T-9, University of California, Berkeley, CA 94720.

J. Spence, R. Helmreich, and J. Stapp, "The Personal Attributes Questionnaire: A Measure of Sex-Role Stereotypes and Masculinity and Femininity," *Journal of Personality and Social Psychology* 32:29-39, July 1975.

S. L. Sutherland, "The Unambitious Female: Women's Low Professional Aspirations," *Signs* 3:774-779, 1978.

E. P. Torrance, "Changing Reactions of Preadolescent Girls to Tasks Requiring Creative Scientific Thinking," *Journal of Genetic Psychology* 102:217-223, 1963.

B. Vetter, "Women in the Natural Sciences," *Signs* 1:713, Spring 1976.

M.S. White, "Psychological and Social Barriers to Women in Science" *Science* 170:413-416, 1970.